# Solut
# tc
# Higher Mathematics



*by*

B. Hastie

ISBN 0 7169 3250 4


The solutions in this publication do not emanate from the Scottish Qualifications Authority. They reflect the author's opinion as to what the solutions might be.

**ROBERT GIBSON · Publisher**
**17 Fitzroy Place, Glasgow, G3 7SF.**

**CONTENTS**

## Model Paper A — Paper 1
(Non-calculator paper)

---

**Question 1**

(a) $\overrightarrow{AB} = \boldsymbol{b} - \boldsymbol{a} = \begin{pmatrix} -1 \\ 3 \\ 2 \end{pmatrix} - \begin{pmatrix} -3 \\ 2 \\ 4 \end{pmatrix} = \underline{\underline{\begin{pmatrix} 2 \\ 1 \\ -2 \end{pmatrix}}}$

(b) $|\overrightarrow{AB}| = \sqrt{2^2 + 1^2 + (-2)^2} = \sqrt{9} = \underline{\underline{3 \text{ units}}}$

---

**Question 2**

(a) If A is the point (–2, 3), then $m_{OA} = \frac{3}{-2}$ $\quad \left\{ m = \frac{y_2 - y_1}{x_2 - x_1} \right\}$

$\Rightarrow m_{NR} = \frac{2}{3}$ $\quad \{ m_1 \times m_2 = -1 \}$

Eqn$_{NR}$ is $y - 3 = \frac{2}{3}(x - (-2))$ $\quad \{ y - b = m(x - a) \}$

$3y - 9 = 2x + 4$

$\underline{\underline{3y = 2x + 13}}$

(b) At B(–5, 1); $\left.\begin{aligned} 3y &= 3(1) &= 3 \\ 2x + 13 &= 2(-5) + 13 &= 3 \end{aligned}\right\}$ $\Rightarrow$ (–5, 1) satisfies the Eqn

$\Rightarrow$ $\underline{\underline{\text{(–5, 1) lies on the line NR}}}$

---

**Question 3**

$g(x) = 3x + 4$

$g(0) = 4 \quad \Rightarrow$ A(0, 4) lies on $f(x)$ and $f(0) = 4$

$f'(x) = 2x - 3$

$f(x) = \int (2x - 3)dx = x^2 - 3x + c$ {Integrate and find $c$}

$f(0) = c = 4$

Hence $f(x) = \underline{\underline{x^2 - 3x + 4}}$

---

**Question 4**

*(a)* $\text{Eqn}_{\text{tangent}}\ y = 2x + 6 \Rightarrow m_{\text{tangent}} = 2 \Rightarrow m_{\text{radius TC}} = \dfrac{-1}{2}$

Centre point C(4, −1) lies on TC $\{m_t \times m_r = -1\}$

$\text{Eqn}_{\text{TC}}$ is $y - (-1) = \dfrac{-1}{2}(x - 4)$ $\{y - b = m(x - a)\}$

$$2y + 2 = -x + 4$$

$$\underline{\underline{x + 2y = 2}}$$

{"touches" ⇒ tangent}

*(b)* $\text{Eqn}_{\text{tan}}$ is $y = 2x + 6 \Rightarrow$ $-2x + y = 6$ Eqn ① × 1 ⇒ $-2x + y = 6$ Eqn ③

$\text{Eqn}_{\text{rad}}$ is $x + 2y = 2$ Eqn ② × 2 ⇒ $2x + 4y = 4$ Eqn ④

③ + ④ ⇒ $5y = 10$

The tangent meets the radius at T(−2, 2) $y = 2$

⇒ $x = -2$

⇒ $\underline{\underline{\text{T}(-2, 2)}}$

By the distance formula

$$\text{TC}^2 = (4 - (-2))^2 + (-1 - 2)^2$$
$$= 36 + 9$$
$$= \underline{\underline{45}}$$

{Use $(x - a)^2 + (y - b)^2 = r^2$}

Hence the equation of the circle is $\underline{\underline{(x - 4)^2 + (y + 1)^2 = 45}}$

**Question 5** $\sin(\text{P} + \text{Q}) = \sin \text{P} \cos \text{Q} + \cos \text{P} \sin \text{Q}$

$$= \frac{12}{13} \cdot \frac{4}{5} + \frac{5}{13} \cdot \frac{3}{5}$$

$$= \frac{48}{65} + \frac{15}{65}$$

$$= \underline{\underline{\frac{63}{65}}}$$

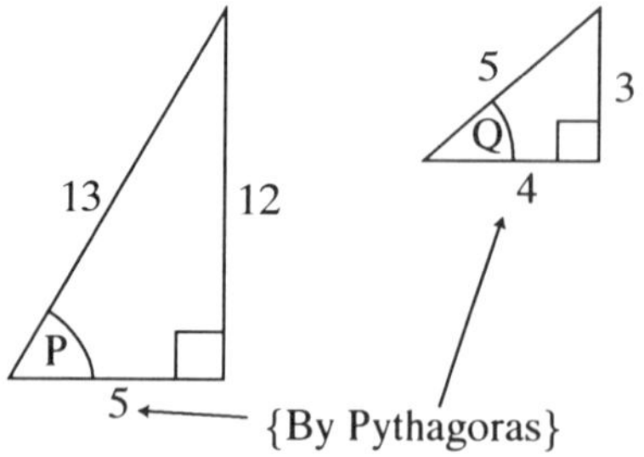

13
12
P
5
5
3
Q
4
{By Pythagoras}

**Question 6** If $x = -3$ is a root, then $f(-3) = 0$

$$\begin{array}{r|rrrr}
 & 2x^3 & -3x^2 & +\ px & +\ 30 \\
-3 & 2 & -3 & p & 30 \\
 & & -6 & 27 & -3p-81 \\
\hline
 & 2 & -9 & p+27 & -3p-51
\end{array}$$

{By synthetic division}

$-3p - 51 = 0 \Rightarrow 3p = -51$

$\underline{\underline{p = -17}}$

$$\begin{array}{r|rrr}
 & 2x^2 & -\ 9x & +\ 10 \\
2 & 2 & -9 & 10 \\
 & & 4 & -10 \\
\hline
 & 2 & -5 & 0
\end{array}$$

$(p + 27 = 10\}$

$\Rightarrow (x-2)$ is a factor $\Rightarrow$ $\underline{\underline{x = 2}}$ is a root

$\Rightarrow (2x-5)$ is a factor $\Rightarrow$ $\underline{\underline{x = \frac{5}{2}}}$ is a root

**Question 7**

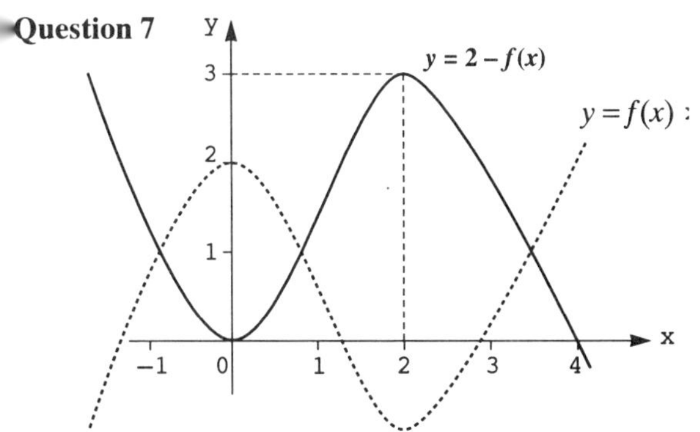


$f(x) \rightarrow -f(x) + 2$

The graph of $y = f(x)$ is given a reflection in $0x$ followed by a translation $\begin{pmatrix} 0 \\ 2 \end{pmatrix}$.

**Question 8**

$$\text{Let } f(x) = 4\sqrt{x} + 3\cos 2x = 4x^{\frac{1}{2}} + 3\cos 2x$$

$$\Rightarrow f'(x) = \frac{1}{2} \,.\, 4x^{\frac{-1}{2}} + 3(-\sin 2x) \,.\, 2$$

$$= \underline{\underline{2x^{\frac{-1}{2}} - 6\sin 2x}} \qquad \left\{\text{or } \frac{2}{\sqrt{x}} - 6\sin 2x\right\}$$

**Question 9**

*(a)* $$\text{LHS} = (\cos x + \sin x)^2 \qquad \left\{\begin{matrix}\cos^2 x + \sin^2 x = 1 \\ 2\sin x\cos x = \sin 2x\end{matrix}\right\}$$

$$= \cos^2 x + 2\sin x\cos x + \sin^2 x$$

$$= \underline{\underline{1 + \sin 2x}} = \text{RHS}$$

*(b)* $$\int (\cos x + \sin x)^2\,dx = \int (1 + \sin 2x)\,dx$$

$$= \underline{\underline{x - \frac{1}{2}\cos 2x + c}}$$

**Question 10** $$\int \sqrt{1+3x}\,dx = \int (1+3x)^{\frac{1}{2}}\,dx$$

$$= \frac{2(1+3x)^{\frac{3}{2}}}{3(3)} + c$$

$$= \underline{\underline{\frac{2}{9}(1+3x)^{\frac{3}{2}} + c}}$$

Hence $$\int_0^1 \sqrt{1+3x}\,dx = \left[\frac{2}{9}(1+3x)^{\frac{3}{2}}\right]_0^1$$

$$= \frac{2}{9}\left((4)^{\frac{3}{2}} - (1)^{\frac{3}{2}}\right) \qquad \left\{4^{\frac{3}{2}} = 8;\ 1^{\frac{3}{2}} = 1\right\}$$

$$= \frac{2}{9}(7)$$

$$= \underline{\underline{\frac{14}{9}}}$$

**Question 11** At $P(p,\ k)$, $k = \log_e p$; At $Q(q,\ k)$, $k = \frac{1}{2}\log_e q$

Hence $$\log_e p = \frac{1}{2}\log_e q \qquad \{x\log a = \log a^x\}$$

$$\log_e p = \log_e q^{\frac{1}{2}}$$

$$\Rightarrow \quad p = q^{\frac{1}{2}}$$

$$\Rightarrow \quad \underline{\underline{p = \sqrt{q}}} \text{ or } \underline{\underline{q = p^2}}$$

If $p = 5$ then $q = 5^2 = \underline{\underline{25}}$

**Question 1**

$$f(x) = x^4 - 2x^3 + 2x - 1 \quad \{f(x) = (x+1)(x-1)^3\}$$

$$f'(x) = 4x^3 - 6x^2 + 2 = 0 \text{ at St. Val.} \quad \{\text{For St. Vs put } f'(x) = 0\}$$

$$2x^3 - 3x^2 + 1 = 0$$

{Use synthetic division to factorise $f'(x)$}

$$\begin{array}{r|rrrr} 1 & 2 & -3 & 0 & 1 \\ & & 2 & -1 & -1 \\ \hline 1 & 2 & -1 & -1 & 0 \\ & & 2 & 1 & \\ \hline & 2 & 1 & 0 & \end{array}$$

$\Rightarrow (x-1)$ is a factor

$\Rightarrow (x-1)$ is a factor

$\Rightarrow (2x+1)$ is a factor

Hence $f'(x) = (2x+1)(x-1)(x-1) = 0$ at St. Val.

$2x+1 = 0$ or $x - 1 = 0$

$\Rightarrow x = -\dfrac{1}{2}$ $\qquad \Rightarrow x = 1$ (twice)

$y = f\left(-\dfrac{1}{2}\right) = \dfrac{-27}{16} \doteqdot -1 \cdot 7$ $\qquad y = f(1) = 0$

Table of Values:

| $x$ | $\frac{-1}{2}^-$ | $\frac{-1}{2}$ | $\frac{-1}{2}^+$ | $1^-$ | $1$ | $1^+$ |
|---|---|---|---|---|---|---|
| $f'(x)$ | − | 0 | + | + | 0 | + |
| shape | ↘ | → | ↗ | ↗ | → | ↗ |
| | | Min. T.P. | | | Pt of Inf. | |

$\left(\dfrac{-1}{2}, \dfrac{-27}{16}\right)$ $\qquad (1,\ 0)$

or $(-0 \cdot 5,\ -1 \cdot 7)$

{Diagram optional}

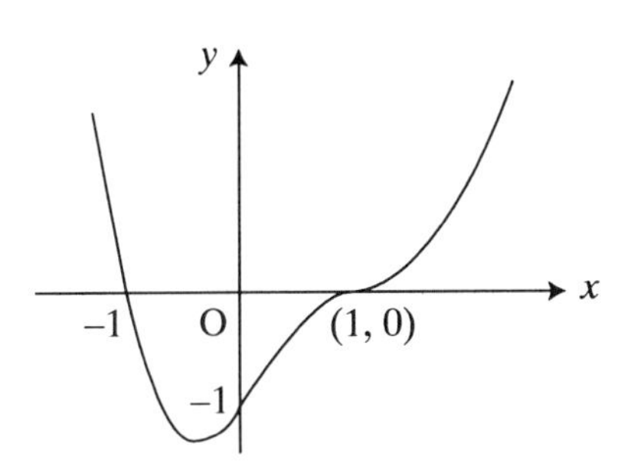


Minimum Turning Point at $\left(\dfrac{-1}{2}, \dfrac{-27}{16}\right)$ $\qquad$ Rising Horizontal Point of Inflexion at (1, 0)

**Question 2**

*(a)* Let $y = 18 - \dfrac{1}{8}x^2 = 0$ on the $x$ axis

$$-\frac{1}{8}x^2 = -18$$

$$x^2 = 144$$

$$x = \pm 12 \Rightarrow \text{A}(-12,\ 0) \text{ and } \text{B}(12, 0)$$

*(b)* Area of Rectangle $= 28 \times 20 = \underline{\underline{560 \text{ ft}^2}}$

Area of Parabola $= \int_{-12}^{12} \left(18 - \frac{1}{8}x^2\right) dx = 2\int_{0}^{12} \left(18 - \frac{1}{8}x^2\right) dx$

$$= 2\left[18x - \frac{x^3}{24}\right]_0^{12}$$

$$= 2\left(216 - \frac{1728}{24}\right)$$

$$= \underline{\underline{288 \text{ ft}^2}}$$

Required area = Area of rectangle − Area of parabola

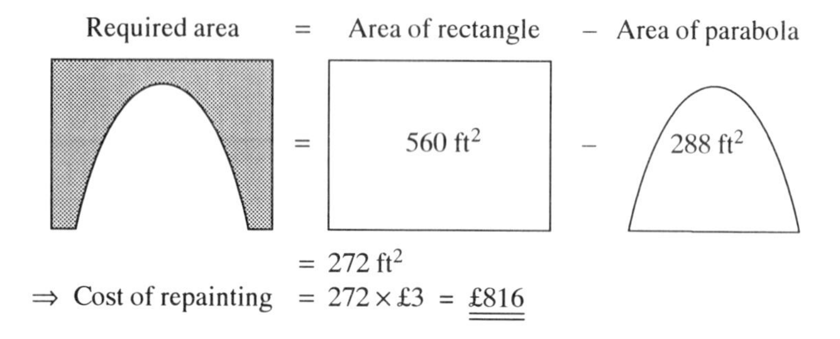


$= 272 \text{ ft}^2$

$\Rightarrow$ Cost of repainting $= 272 \times £3 = \underline{\underline{£816}}$

---

**Question 3**

*(a)* $k(x) = f(g(x)) = f(3-2x) = 2(3-2x) - 1 = 6 - 4x - 1 = \underline{\underline{5-4x}}$

*(b)* $h(k(x)) = h(5-4x) = \frac{1}{4}(5-(5-4x)) = \frac{1}{4}(4x) = \underline{\underline{x}}$

*(c)* Since $h(k(x)) = x$

Then $\underline{\underline{h = k^{-1}(x)}}$ i.e. $h$ and $k$ are <u>INVERSES</u> of each other

---

**Question 4**

*a)* The line (road) meets the curve (circuit) where:

$-4x - 3 = 5 - 2x^2 - x^3$

$x^3 + 2x^2 - 4x - 8 = 0$

$(x+2)(x+2)(x-2) = 0$

$x = -2$ (twice) and $x = 2$

$\Rightarrow$ tangent at $x = -2$

where $y = 5$

Factorise $f(x) = x^3 + 2x^2 - 4x - 8$

| | | | | |
|---|---|---|---|---|
| $-2$ | 1 | 2 | $-4$ | $-8$ |
| | | $-2$ | 0 | 8 |
| $-2$ | 1 | 0 | $-4$ | $0 \Rightarrow (x+2)$ is a factor |
| | | $-2$ | 4 | |
| | 1 | $-2$ | $0 \Rightarrow (x+2)$ is a factor | |

$\Rightarrow (x-2)$ is a factor

Hence $\underline{\underline{B(-2,\ 5)}}$ is the point of contact

*b)* At B(−2, 5) $\{f(-2) = 5\}$

gradient of line $y = -4x - 3 = \underline{\underline{-4}}$

gradient of curve $y = 5 - 2x^2 - x^3$ $\quad \dfrac{dy}{dx} = -4x - 3x^2$ $\quad \{y = mx + c\}$

At $x = -2$ $\quad \dfrac{dy}{dx} = 8 - 12 = \underline{\underline{-4}}$ $\quad \left\{m = \dfrac{dy}{dx}\right\}$

Since the $\underline{\underline{\text{2 gradients are equal}}}$, then the drivers go straight on

**Question 5**

*a)* "15% lost" $\Rightarrow$ 85% remains each hour $\{85\% = 0{\cdot}85\}$

$\Rightarrow$ After 4 hours, the amount of serum remaining $= 0{\cdot}85^4 \times 25$

$\doteqdot \underline{\underline{13{\cdot}05 \text{ mg}}}$

*b)* After 1 dose, amount $\doteqdot 13{\cdot}05$ mg $< 20$

After 2 doses, amount $= (13{\cdot}05 + 25) \times 0{\cdot}85^4 \doteqdot 19{\cdot}86$ mg $< 20$

After 3 doses, amount $= (19{\cdot}86 + 25) \times 0{\cdot}85^4 \doteqdot 23{\cdot}42$ mg $> 20$

$\underline{\underline{\text{Hence 3 doses are needed}}}$

*c)* $u_{n+1} = 0{\cdot}85^4\, u_n + 25$

$\Rightarrow u_{n+1} = \underline{\underline{0{\cdot}522\, u_n + 25}}$ $\quad$ since $0{\cdot}85^4 = 0{\cdot}522$

*d)* From *(c)* $\quad u_{n+1} = 0{\cdot}522\, u_n + 25 \quad \Rightarrow$

$u_1 = 25$

$u_2 = 38{\cdot}05$

$u_3 = 44{\cdot}86$

$u_4 = 48{\cdot}42$

$u_5 = 50{\cdot}27$

$\vdots \quad \vdots \quad \vdots$

$u_{10} = 52{\cdot}2$

The level seems to approach a LIMIT

If this limit is $L$

Then $\quad L = 0{\cdot}522\, L + 25$

$0{\cdot}478\, L = 25$

$L \doteqdot 52{\cdot}3$ mg $\quad \Rightarrow \quad \underline{\underline{\text{No maximum length of time}}}$

{Since $52{\cdot}3 < 55$}

**Question 6**

*(a)* Required area $A$ = area $\Delta$ + area of rectangle ACDE

Area $\Delta$ABC $= \frac{1}{2} \times 4 \times 4 \sin \theta° = \underline{\underline{8 \sin \theta°}}$

Area ACDE $= \frac{1}{2}$ (area square ACFG)

$= \frac{1}{2}\ AC^2$

$= \frac{1}{2}\ (4^2 + 4^2 - 2 \times 4 \times 4 \cos\ \theta°)$

$= \frac{1}{2}\ (32 - 32 \cos\ \theta°)$

$= \underline{\underline{16 - 16 \cos\ \theta°}}$

{Use Cosine Rule to find $AC^2$}

or alternatively use

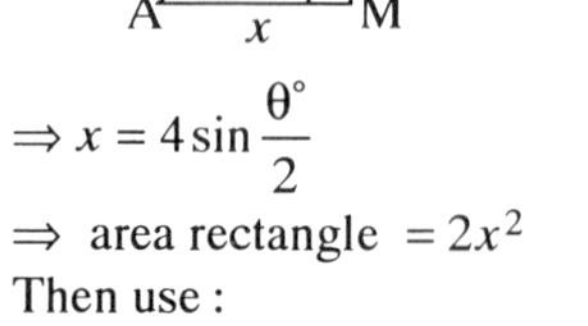


$\Rightarrow x = 4 \sin \frac{\theta°}{2}$

$\Rightarrow$ area rectangle $= 2x^2$

Then use :

$\sin^2 \frac{\theta°}{2} = \frac{1}{2}(1 - \cos\theta°)$ etc.

Hence $A$ $= 8\sin\ \theta° + 16 - 16\cos\ \theta°$

$= \underline{\underline{8(2 + \sin\ \theta° - 2\cos\ \theta°)}}$

*(b)* $8\sin\ \theta° - 16\cos\ \theta° \quad = k\sin(\theta - \alpha)°$

$= k\sin\ \theta° \cos\ \alpha° - k\cos\ \theta°\ \sin\ \alpha°$

Here $\left.\begin{aligned} k\cos\ \alpha° &= 8 \\ k\sin\ \alpha° &= 16 \end{aligned}\right\} \Rightarrow k^2 = 8^2 + 16^2 = 320$

$\Rightarrow \quad k = \sqrt{320}$

$= \underline{\underline{8\sqrt{5}}}$

$\Rightarrow \tan\ \alpha° \quad = \frac{16}{8} = 2$

$\Rightarrow \underline{\underline{\alpha = 63 \cdot 4}}$

Hence $\underline{\underline{8\sin\ \theta° - 16\cos\ \theta° = 8\sqrt{5}\sin(\theta - 63 \cdot 4)°}}$

*(c)* When $A = 8\sin\ \theta° + 16 - 16\cos\ \theta° = 30$ {From part *(a)*}

Then $16 + 8\sqrt{5}\sin(\theta - 63 \cdot 4)° = 30$ {From part *(b)*}

$8\sqrt{5}\sin(\theta - 63 \cdot 4)° = 14$

$\sin(\theta - 63 \cdot 4)° \doteqdot \frac{14}{8\sqrt{5}}$ $\{ \doteqdot 0 \cdot 7826\}$

$\theta - 63 \cdot 4 = 51 \cdot 5$ {Ignore $128 \cdot 5$ as $\theta < 180$}

$\underline{\underline{\theta = 114 \cdot 9}}$

**Question 7**

*(a)* Since the graph in figure 2 is LINEAR, then $\log_e I = m\log_e t + c$

Here, $\underline{\underline{c = 4}}$ since (0, 4) is on the line

Also, $m = \dfrac{4-0}{0-5} = \underline{\underline{\dfrac{-4}{5}}}$ $\Rightarrow$ Eqn is $\underline{\underline{\log_e I = \dfrac{-4}{5}\log_e t + 4}}$

*(b)* If $I = kt^r$

Then $\log_e I = \log_e(kt^r) = \log_e t^r + \log_e k$ {Take logs both sides}

$\Rightarrow \log_e I = r\log_e t + \log_e k$ {Compare with *(a)*}

From *(a)* $r = \dfrac{-4}{5} = \underline{\underline{-0\cdot 8}}$ Also $\log_e k = 4$

$\Rightarrow k = e^4 \doteqdot \underline{\underline{54\cdot 6}}$

Hence $\underline{\underline{I = 54\cdot 6t^{-0\cdot 8}}}$

**Question 8**

*(a)*

$$C(x) = 2x + y$$
$$= 2x + 100 - d$$
$$= 2x + 100 - \sqrt{x^2 - 243}$$
$$\Rightarrow \underline{\underline{C(x) = 2x + 100 - (x^2 - 243)^{\frac{1}{2}}}}$$

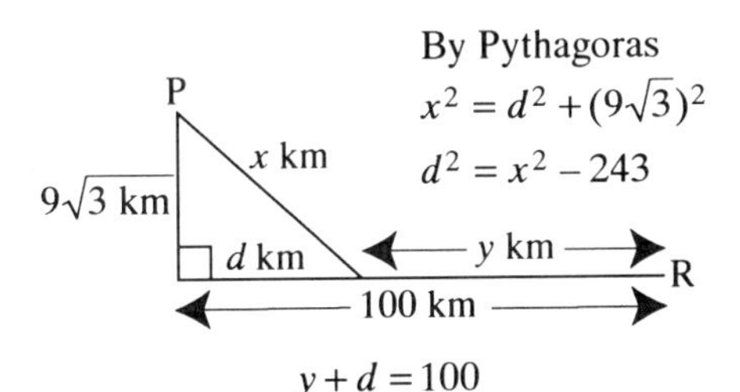


By Pythagoras

$x^2 = d^2 + (9\sqrt{3})^2$

$d^2 = x^2 - 243$

$y + d = 100$

$\underline{\underline{y = 100 - d}}$

*(b)* From *(a)*

$$C(x) = 2x + 100 - (x^2 - 243)^{\frac{1}{2}}$$
$$\Rightarrow C_1(x) = 2 - \frac{1}{2}(x^2 - 243)^{\frac{-1}{2}}(2x) = 0 \text{ at St.Val.}$$
$$\frac{x}{(x^2-243)^{\frac{1}{2}}} = 2$$
$$\frac{x^2}{x^2-243} = 4$$
$$x^2 = 4x^2 - 972$$
$$972 = 3x^2$$
$$x^2 = 324$$
$$\underline{\underline{x = 18}}$$
$$\underline{\underline{C(18) = 127}}$$

{Ignore $x = -18$}

Table of Values

| $x$ | $18^-$ | 18 | $18^+$ |
|---|---|---|---|
| $C'(x)$ | – | 0 | + |
| shape | ↘ | → | ↗ |
| | | | |

Min at $\underline{\underline{(18, 127)}}$

Hence minimum cost = $\underline{\underline{£127 \text{ million}}}$

Total length $= x + y = 18 + 100 - (18^2 - 243)^{\frac{1}{2}}$

$= 118 - (81)^{\frac{1}{2}}$

$\Rightarrow$ Total length = $\underline{\underline{109 \text{ km}}}$ {or 127 – 18}

## Model Paper B — Paper 1

(Non-calculator paper)

---

**Question 1**

$$\int(3x^3+4x)dx = \frac{3x^4}{4}+\frac{4x^2}{2}+c = \underline{\underline{\frac{3}{4}x^4+2x^2+c}}$$

---

**Question 2**

$$\vec{RS} = \boldsymbol{s}-\boldsymbol{r} = \begin{pmatrix}2\\-5\\4\end{pmatrix}-\begin{pmatrix}-1\\-8\\-2\end{pmatrix}=\begin{pmatrix}3\\3\\6\end{pmatrix}$$

$$\vec{ST} = \boldsymbol{t}-\boldsymbol{s} = \begin{pmatrix}3\\-4\\6\end{pmatrix}-\begin{pmatrix}2\\-5\\4\end{pmatrix}=\begin{pmatrix}1\\1\\2\end{pmatrix}$$

Since $\vec{RS}=3\vec{ST}$, then $\underline{\underline{\text{R, S and T are collinear}}}$

---

**Question 3**

*(a)*

$$\begin{aligned} f(x) &= 4x^2-3x+5 \\ \Rightarrow f(x+1) &= 4(x+1)^2-3(x+1)+5 \\ &= 4(x^2+2x+1)-3x-3+5 \\ &= \underline{\underline{4x^2+5x+6}} \end{aligned}$$

$$\begin{aligned} f(x-1) &= 4(x-1)^2-3(x-1)+5 \\ &= 4(x^2-2x+1)-3x+3+5 \\ &= \underline{\underline{4x^2-11x+12}} \end{aligned}$$

$$\begin{aligned} \frac{f(x+1)-f(x-1)}{2} &= \frac{4x^2+5x+6-(4x^2-11x+12)}{2} \\ &= \frac{16x-6}{2} \\ &= \underline{\underline{8x-3}} \end{aligned}$$

*(b)*

$$\begin{aligned} g(x) &= 2x^2 + 7x - 8 \\ g(x+1) &= 2(x+1)^2 + 7(x+1) - 8 \\ &= 2(x^2+2x+1) + 7x + 7 - 8 = 2x^2 + 11x + 1 \\ g(x-1) &= 2(x-1)^2 + 7(x-1) - 8 \\ &= 2(x^2-2x+1) + 7x - 7 - 8 = 2x^2 + 3x - 13 \end{aligned}$$

Hence $\dfrac{g(x+1)-g(x-1)}{2} = \dfrac{2x^2+11x+1-(2x^2+3x-13)}{2}$

$$= \frac{8x+14}{2}$$

$$= \underline{\underline{4x+7}}$$

*(c)*

$f(x) = 4x^2 - 3x + 5;\ \dfrac{f(x+1)-f(x-1)}{2} = 8x - 3;\ \{f'(x) = 8x - 3\}$

$g(x) = 2x^2 + 7x - 8;\ \dfrac{g(x+1)-g(x-1)}{2} = 4x + 7;\ \{g'(x) = 4x + 7\}$

If $h(x) = 3x^2 + 5x - 1$

Then $\dfrac{h(x+1)-h(x-1)}{2} = \underline{\underline{6x+5}}$ as this is $h'(x)$

---

**Question 4**

$$\boldsymbol{u}+\boldsymbol{v} = \begin{pmatrix} -3 \\ 3 \\ 3 \end{pmatrix} + \begin{pmatrix} 1 \\ 5 \\ -1 \end{pmatrix} = \begin{pmatrix} -2 \\ 8 \\ 2 \end{pmatrix}$$

$$\boldsymbol{u}-\boldsymbol{v} = \begin{pmatrix} -3 \\ 3 \\ 3 \end{pmatrix} - \begin{pmatrix} 1 \\ 5 \\ -1 \end{pmatrix} = \begin{pmatrix} -4 \\ -2 \\ 4 \end{pmatrix}$$

$$(\boldsymbol{u}+\boldsymbol{v}) \cdot (\boldsymbol{u}-\boldsymbol{v}) = \begin{pmatrix} -2 \\ 8 \\ 2 \end{pmatrix} \cdot \begin{pmatrix} -4 \\ -2 \\ 4 \end{pmatrix} = 8 - 16 + 8 = 0$$

Since $(\boldsymbol{u}+\boldsymbol{v}) \cdot (\boldsymbol{u}-\boldsymbol{v}) = 0$, then $\underline{\underline{(\boldsymbol{u}+\boldsymbol{v}) \text{ and } (\boldsymbol{u}-\boldsymbol{v}) \text{ are perpendicular}}}$

---

**Question 5**

*(a)* By substituting $y = x$ into $x^2 + y^2 - 6x - 2y - 24 = 0$

We have

$$\begin{aligned} x^2 + x^2 - 6x - 2x - 24 &= 0 \\ 2x^2 - 8x - 24 &= 0 \\ x^2 - 4x - 12 &= 0 \\ (x+2)(x-6) &= 0 \\ x + 2 = 0 \quad \text{or} \quad x - 6 &= 0 \\ x = -2 \,; \quad x &= 6 \\ y = -2 \quad y &= 6 \end{aligned}$$

Hence A(6, 6) and B(−2, −2)

*(b)* Let C be the mid - point of AB.

Then C, the point (2, 2), is the circle centre and AC is a radius

$AC^2 = (2-(-2))^2 + (2-(-2))^2 = 32$ {using the distance formula}

Hence the equation of the circle on AB is $(x-2)^2 + (y-2)^2 = 32$

{or $x^2 + y^2 - 4x - 4y - 24 = 0$}

**Question 6**

*(a)* $u_2 = 0 \cdot 9u_1 + 2 = 0 \cdot 9(3) + 2$

Hence $u_2 = 4 \cdot 7$

*(b)*

$$\begin{aligned} u_3 &= 0{\cdot}9u_2 + 2 \\ &= 0{\cdot}9 \times 4{\cdot}7 + 2 \\ \Rightarrow u_3 &= 6{\cdot}23 > 6 \end{aligned}$$

Hence the required value is $n = 3$

*(c)* At the limit, let $u_n = u_{n+1} = u$

$$\begin{aligned} \text{Then } u &= 0 \cdot 9u + 2 \\ 0 \cdot 1u &= 2 \\ u &= 20 \end{aligned}$$

Hence the limit is 20

**Question 7**

$$\begin{aligned} x^2+8x+18 &= x^2+8x+16-16+18 \\ &= x^2+8x+16+2 \\ &= \underline{\underline{(x+4)^2+2}} \end{aligned}$$

Hence minimum turning point at $\underline{\underline{(-4,\ 2)}}$

**Question 8**

*(a)*

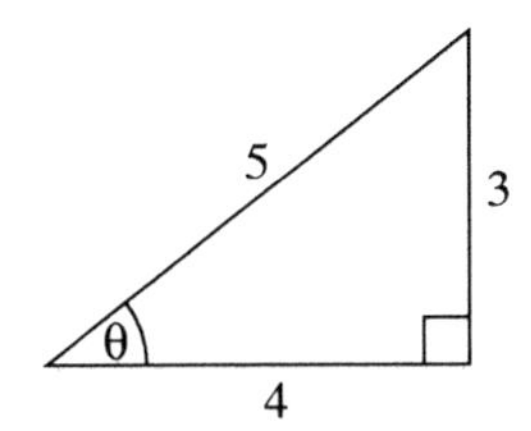


{By Pythagoras}

$$\begin{aligned} \sin 2\theta &= 2\sin\theta\cos\theta \\ &= 2\,.\,\frac{3}{5}\,.\,\frac{4}{5} = \underline{\underline{\frac{24}{25}}} \end{aligned}$$

Also
$$\begin{aligned} \cos 2\theta &= \cos^2\theta - \sin^2\theta \\ &= \left(\frac{4}{5}\right)^2 - \left(\frac{3}{5}\right)^2 = \frac{7}{25} \end{aligned}$$

*(b)*
$$\begin{aligned} \sin 4\theta &= 2\sin 2\theta\cos 2\theta \\ &= 2\left(\frac{24}{25}\right)\cdot\left(\frac{7}{25}\right) \\ &= \underline{\underline{\frac{336}{625}}} \end{aligned}$$

**Question 9**

$$y = f(x) = \sqrt{3}x - x^2$$

$$f'(x) = \sqrt{3} - 2x$$

$$f'(0) = \sqrt{3} \Rightarrow \underline{m = \sqrt{3}} \qquad \{m = f'(x)\}$$

Also $\tan\theta° = \sqrt{3}$ $\qquad \{m = \tan\theta°\}$

$\Rightarrow \theta = 60$

For $y = x$; $\tan\alpha° = 1$ $\qquad \{m = \tan\alpha\}$

$\Rightarrow \alpha = 45°$

Hence the required angle is $\underline{\underline{15°}}$ $\qquad \{60° - 45°\}$

**Question 10**

*(a)*

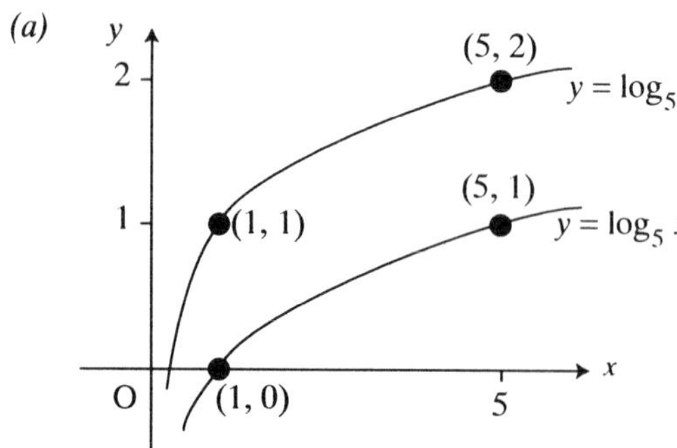


Cuts $x$ - axis where

$$\begin{aligned} \log_5 x + 1 &= 0 \\ \log_5 x &= -1 \\ x &= 5^{-1} \\ x &= \underline{\underline{\frac{1}{5} \text{ or } 0 \cdot 2}} \end{aligned}$$

Hence the crossing point is $\underline{\underline{\left(\frac{1}{5}, \ 0\right)}}$

*(b)*

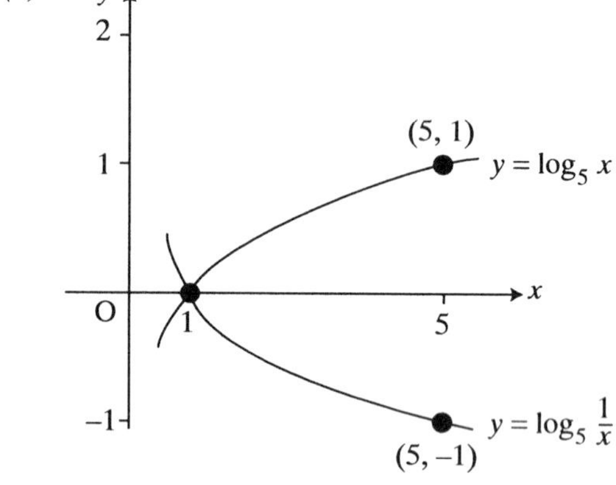


***N.B.***

$$\begin{aligned} y &= \log_5 \frac{1}{x} \\ &= \log_5 x^{-1} \\ &= \underline{\underline{-\log_5 x}} \end{aligned}$$

{Reflection of $y = \log_5 x$}

**Question 11**

$$\begin{aligned} \text{Let } f(x) &= \sin^3 x = (\sin x)^3 \\ \Rightarrow f'(x) &= 3(\sin x)^2 \,.\, \cos x \\ &= \underline{\underline{3\sin^2 x \cos x}} \end{aligned}$$

$$\begin{aligned} \int \sin^2 x \cos x \, dx &= \frac{1}{3}\int 3\sin^2 x \cos x \, dx \\ &= \underline{\underline{\frac{1}{3}\sin^3 x + c}} \end{aligned}$$

{Inverse process}

**Question 1**

*(a)* $f(x) = 2x^3 + x^2 - 13x + a$ $\{y - \text{intercept} = f(0) = a\}$

Since $x = 2$ is a root then $f(2) = 0$

$$\begin{array}{c|cccc} 2 & 2 & 1 & -13 & a \\ & & 4 & 10 & -6 \\ \hline -3 & 2 & 5 & -3 & 0 \\ & & -6 & 3 & \\ \hline & 2 & -1 & 0 & \end{array}$$

$\Rightarrow a - 6 = 0$

$\underline{\underline{a = 6}} \Rightarrow \underline{\underline{\text{point } (0, 6)}}$

*(b)* $0 \Rightarrow (x + 3)$ is a factor

$\underline{x = -3 \text{ is a root}}$

$\Rightarrow (2x - 1)$ is a factor

$\underline{x = \frac{1}{2} \text{ is a root}}$

Hence other points are $\underline{\underline{(-3,\ 0) \text{ and } \left(\frac{1}{2},\ 0\right)}}$

**Question 2**

*(a)* $m_{OD} = \frac{1}{2} \Rightarrow m_{AD} = -2$ $\{m_1 \times m_2 = -1\}$

$\text{Equation}_{AD} \begin{cases} \text{point A}(3,\ 4) \\ \text{gradient} = -2 \end{cases}$ $\{\text{use } y - b = m(x - a)\}$

$y - 4 = -2(x - 3)$

$\underline{\underline{y = -2x + 10}}$

*(b)* At D, $y = \frac{1}{2}x$ {Simultaneous equations}

and $\underline{y = -2x + 10}$

Hence $0 = 2\frac{1}{2}x - 10$

By substitution $\left.\begin{matrix} x = 4 \\ y = 2 \end{matrix}\right\} \Rightarrow \underline{\underline{\text{D is the point } (4,\ 2)}}$

*(c)* Area ABCD $= AD^2 = (4 - 3)^2 + (2 - 4)^2$ {Distance formula}

$\Rightarrow$ Area $= \underline{\underline{5 \text{ units}^2}}$

## Question 3

*(a)* B(6, 4, 2); C(4, 3, 4); D(6, 2, 2)

*(b)* Mid point of AD $= \left(\frac{2+6}{2}, \frac{4+2}{2}, \frac{6+2}{2}\right)$

$= \underline{\underline{(4,\ 3,\ 4) = \mathrm{C}}}$

*(c)* $\vec{OA} = \boldsymbol{a} = \begin{pmatrix} 2 \\ 4 \\ 6 \end{pmatrix} \Rightarrow |\boldsymbol{a}| = \sqrt{2^2+4^2+6^2} = \sqrt{56}$

$\vec{OB} = \boldsymbol{b} = \begin{pmatrix} 6 \\ 4 \\ 2 \end{pmatrix} \Rightarrow |\boldsymbol{b}| = \sqrt{6^2+4^2+2^2} = \sqrt{56}$

$$\cos A\hat{O}B = \frac{\boldsymbol{a}\,.\,\boldsymbol{b}}{|\boldsymbol{a}||\boldsymbol{b}|} = \frac{\begin{pmatrix} 2 \\ 4 \\ 6 \end{pmatrix} . \begin{pmatrix} 6 \\ 4 \\ 2 \end{pmatrix}}{\sqrt{56}\sqrt{56}} = \frac{12+16+12}{56} = \frac{40}{56}$$

$\Rightarrow A\hat{O}B = \underline{\underline{44 \cdot 4^\circ}}$

*(d)* Since $\mathrm{OA} = \mathrm{OB} = \sqrt{56}$

Then ΔABO is isosceles $\Rightarrow$ $2A^\circ + 44\cdot4^\circ = 180^\circ$

$2A^\circ = 135\cdot6^\circ$

Hence angle OAB $= \underline{\underline{67\cdot8^\circ}}$

O 44·4° A°

## Question 4

*(a)* (i) From the equations of the circles:

Centre of small wheel is (0, 3)
Centre of large wheel is (14, 10)

distance$^2 = 14^2 + 7^2$

$= 245$

$\Rightarrow$ distance $= \sqrt{245}$

$\doteqdot 15\cdot65$ units

as 1 unit = 5 cm then distance $\doteqdot \underline{\underline{78\cdot3 \text{ cm}}}$

(ii) $\text{Radius}_{\text{small}} = \sqrt{0+9-0} = 3$

$\text{Radius}_{\text{large}} = \sqrt{14^2+10^2-196} = 10$

sum of radii = 13 units

$= \underline{\underline{65 \text{ cm}}}$

Hence the clearance is $78\cdot3 - 65 = 13\cdot3$ cm

$= \underline{\underline{133 \text{ mm}}}$

*(b)* (i) B(7, 3), P(14,10) $\Rightarrow$ $m_{BP} = \dfrac{10-3}{14-7} = \underline{\underline{1}}$

(ii) Since BP bisects AC, then BP is perpendicular to AC

Hence $m_{AC} = -1$ $\qquad \{m_1 \times m_2 = -1\}$

$$\text{Equation}_{AC} \begin{cases} B(7,\ 3) \\ \qquad y-3 = -1(x-7) \\ \qquad\quad y = -x+10 \\ m=-1 \end{cases}$$

Eqn chord $\quad y = -x+10 \qquad \left\{\begin{array}{rl} y^2 = & x^2-20x+100 \\ -20y = & 20x-200 \end{array}\right\}$

Eqn circle $\quad x^2+y^2-28x-20y+196 = 0$

By substitution :

$$\begin{aligned} x^2+x^2-20x+100-28x+20x-200+196 &= 0 \\ 2x^2-28x+96 &= 0 \\ x^2-14x+48 &= 0 \\ (x-8)(x-6) &= 0 \\ x = 8 \text{ or } x &= 6 \\ \underline{y = 2} \qquad \underline{y = 4} \end{aligned}$$

Hence $\underline{\underline{A(8,\ 2) \text{ and } C(6,\ 4)}}$

---

**Question 5**

*(a)*

$$\begin{aligned} f(x) = 3\sin x^\circ - \cos x^\circ &= k\sin\ (x-\alpha)^\circ \\ 3\sin x^\circ - \cos x^\circ &= k\sin\ x^\circ\cos\alpha^\circ - k\cos\ x^\circ\sin\alpha^\circ \end{aligned}$$

Hence $\left.\begin{array}{r} -k\sin\alpha^\circ = -1 \\ k\cos\alpha^\circ = 3 \end{array}\right\} \Rightarrow -\tan\ \alpha^\circ = \dfrac{-1}{3}$

$\underline{\alpha = 18\cdot 4}$

Also $\quad k^2 = (-1)^2+3^2 = 10$

$\underline{\underline{k = \sqrt{10}}}$

Hence $\quad f(x) = \underline{\underline{3\sin x^\circ - \cos x^\circ = \sqrt{10}\sin(x-18\cdot 4)^\circ}}$

*(b)*

$$\begin{aligned} 3\sin x^\circ - \cos x^\circ &= \sqrt{5} \\ \Rightarrow \sqrt{10}\sin(x - 18\cdot 4)^\circ &= \sqrt{5} \quad \text{\{From part (a)\}} \\ \sin(x - 18\cdot 4)^\circ &= \frac{\sqrt{5}}{\sqrt{10}} \\ \sin(x - 18\cdot 4)^\circ &= 0\cdot 707 \\ x - 18\cdot 4 &= 45,\ 135 \\ x &= \underline{\underline{63\cdot 4,\ 153\cdot 4}} \end{aligned}$$

---

**Question 6**

*(a)*

$$\begin{aligned} m_t = m_0 e^{-0\cdot 02t} \Rightarrow m_{10} &= 500e^{-0\cdot 02\times 10} \\ &= \underline{\underline{409\cdot 37\text{ g}}} \end{aligned}$$

*(b)*

$$\begin{aligned} m = 500e^{-0\cdot 02t} &= 250 \\ e^{-0\cdot 02t} &= 0\cdot 5 \quad \text{\{Take } \log_e \text{ of both sides\}} \\ -0\cdot 02t \ln e &= \ln 0\cdot 5 \\ t &= \frac{\ln 0\cdot 5}{-0\cdot 02} \quad \{\ln = \log_e\} \\ t &= 34\cdot 66 \end{aligned}$$

Hence the half - life $\doteqdot$ $\underline{\underline{34\cdot 7\text{ years}}}$

*(c)*

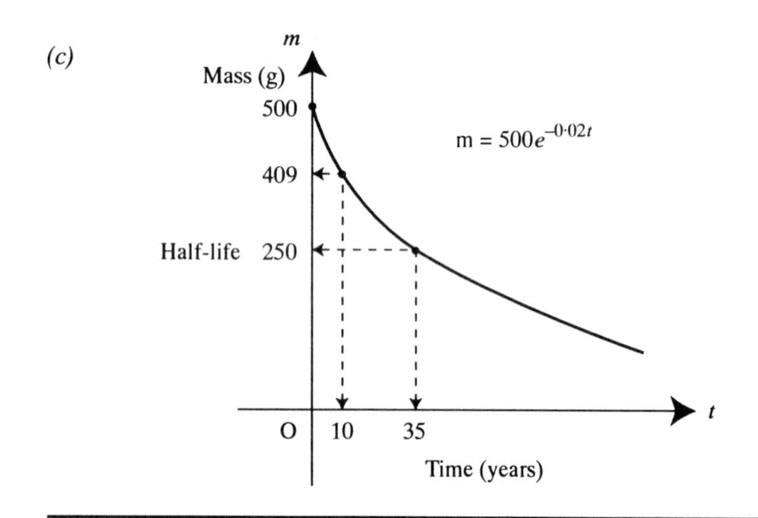


---

**Question 7**

*(a)*

$$\text{Area } \Delta\text{AEF} = \text{area rectangle ABCD} - (\text{area of } \Delta\text{s ABE} + \text{ADF} + \text{ECF})$$

$$= 8 \times 6 - \left(\frac{1}{2} \times 8x + \frac{1}{2}(8-x)6 + \frac{1}{2}(6-x)x\right)$$

$$= 48 - \left(4x + 24 - 3x + 3x - \frac{x^2}{2}\right)$$

$$\Rightarrow H(x) = \underline{\underline{24 - 4x + \frac{x^2}{2}}}$$

*(b)*

$$H(x) = \frac{x^2}{2} - 4x + 24$$

$$H'(x) = x - 4 = \text{at St. Val.}$$

$$x = 4$$

$$y = 16$$

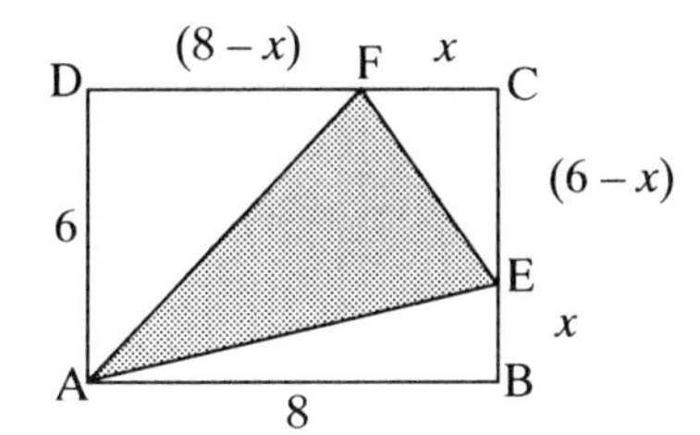


| $x$ | $4^-$ | $4$ | $4^+$ |
|---|---|---|---|
| $H'(x)$ | – | 0 | + |
| shape | ↘ | → | ↗ |
| | | Min T.P. at (4, 16) | |

Also, if $x = 0$
then $H(0) = 24$
This is the maximum area

Hence least area $= 16$ units$^2$
greatest area $= 24$ units$^2$

**Question 8**

*(a)*

$$\begin{aligned} y &= x^2 + px + q \\ \text{At A(2, 2);} \quad 2 &= 2^2 + 2p + q \\ -2 &= 2p + q \\ q &= \underline{\underline{-2p-2}} \quad \text{or} \quad p = \underline{\underline{-\frac{1}{2}(q+2)}} \end{aligned}$$

*(b)*

$$\begin{aligned} f(x) &= x^2 + px + q \\ f'(x) &= 2x + p = \text{gradient of the curve} \\ f'(2) &= 4 + p = 1 \qquad \{\text{gradient of tangent at A}\} \end{aligned}$$

$$\left.\begin{aligned} p &= -3 \\ \underline{\underline{q}} &\underline{\underline{= 4}} \end{aligned}\right\} \quad \text{Hence Eqn is } \underline{\underline{y = x^2 - 3x + 4}}$$

*(c)*

For $y = x^2 - 3x + 4$;

$$\left.\begin{aligned} a &= 1 \\ b &= -3 \end{aligned}\right\} \Rightarrow b^2 - 4ac = 9 - 4 \times 1 \times 4 = -7, \quad \underline{\underline{\text{no real roots}}}$$

$$c = 4 \qquad \underline{\underline{\text{Curve doesn't meet the } x\text{-axis}}}$$

**Question 9**

At $y = 1$; $\frac{1}{4}x^2 = 1$, $x^2 = 4$, $\underline{\underline{x = \pm 2}}$

At $y = 9$; $\frac{1}{4}x^2 = 9$, $x^2 = 36$, $\underline{\underline{x = \pm 6}}$

This area is symmetrical about the $y$-axis.

Area in the first quadrant can be split up as shown :

$$\text{Area} = 8\times 2 + 9\times 4 - \int_2^6 \frac{1}{4}x^2\,dx$$

$$= 52 - \left[\frac{x^3}{12}\right]_2^6$$

$$= 52 - \left\{\left(\frac{6^3}{12}\right) - \left(\frac{2^3}{12}\right)\right\}$$

$$= 52 - \left\{18 - \frac{2}{3}\right\}$$

$$= 52 - 17\frac{1}{3}$$

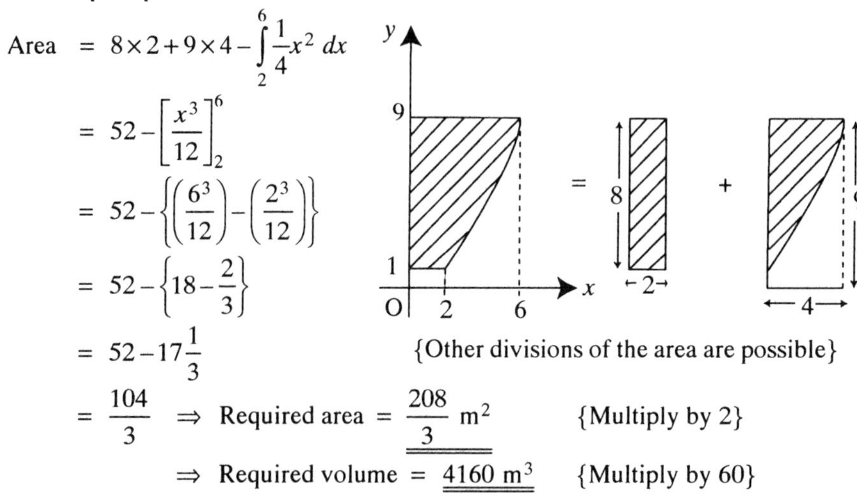


{Other divisions of the area are possible}

$$= \frac{104}{3} \Rightarrow \text{Required area} = \underline{\underline{\frac{208}{3}\text{ m}^2}} \qquad \{\text{Multiply by 2}\}$$

$$\Rightarrow \text{Required volume} = \underline{\underline{4160\text{ m}^3}} \qquad \{\text{Multiply by 60}\}$$

Alternative solution to Q10, involves integration w.r. to "$y$"

If $y = \frac{1}{4}x^2$ then $x = 2\sqrt{y}$

$$= 2y^{\frac{1}{2}}$$

$$\text{Area} = 2\int_1^9 \left(2y^{\frac{1}{2}}\right) dy$$

$$= 2\left[\frac{4}{3}y^{\frac{3}{2}}\right]_1^9$$

$$= 2\left\{\left(\frac{4}{3}\cdot 9^{\frac{3}{2}}\right) - \left(\frac{4}{3}\cdot 1^{\frac{3}{2}}\right)\right\}$$

$$= 2\left(36 - \frac{4}{3}\right)$$

$$= \underline{\underline{69\tfrac{1}{3}\text{ m}^2}} \Rightarrow \text{vol} = \underline{\underline{4160\text{ m}^3}}$$

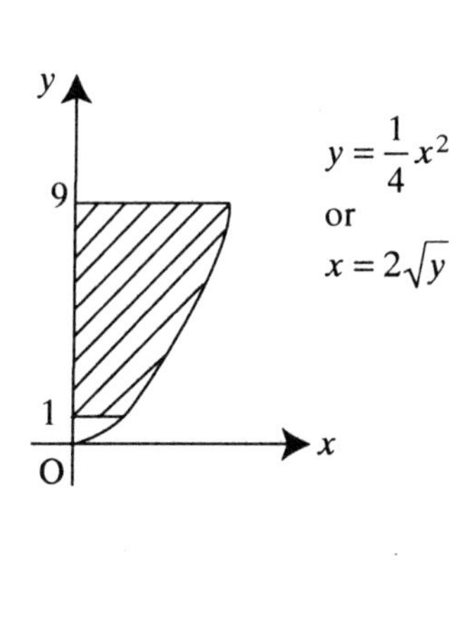

# Model Paper C — Paper 1

(Non-calculator paper)

**Question 1**

*(a)* If $x-3$ is a factor, then $f(3)=0$

$f(x)=2x^3 + 3x^2 - 23x - 12$

$$\begin{array}{r|rrrr} 3 & 2 & 3 & -23 & -12 \\ & & 6 & 27 & 12 \\ \hline & 2 & 9 & 4 & 0 \end{array} \Rightarrow \underline{\underline{(x-3)\text{ is a factor}}}$$

*(b)*

$$\begin{array}{r|rrr} -4 & 2 & 9 & 4 \\ & & -8 & -4 \\ \hline & 2 & 1 & 0 \end{array} \Rightarrow (x+4)\text{ is a factor}$$

Hence $(2x+1)$ is a factor

so $f(x)=2x^3+3x^2-23x-12 \;=\; \underline{\underline{(x-3)(x+4)(2x+1)}}$

**Question 2**

$$\int\left(6x^2-x+\cos x\right)dx \;=\; \frac{6x^3}{3}-\frac{x^2}{2}+\sin x+c$$

$$= \underline{\underline{2x^3-\frac{1}{2}x^2+\sin x+c}}$$

**Question 3**

*(a)* In $\Delta$ABC; A(4, 8); B(1, 2)

$\Rightarrow \; AB^2 = (8-2)^2 + (4-1) = 36 + 9 = 45$

A(4, 8); C(7, 2)

$\Rightarrow \; AC^2 = (8-2)^2 + (4-7)^2 = 36 + 9 = 45$

Since AB = AC = $\sqrt{45}$ $\{3\sqrt{5}\}$

Then $\underline{\underline{\Delta ABC \text{ is isosceles}}}$

*(b)* (i) $m_{BC} = \dfrac{2-2}{7-1} = 0 \Rightarrow m_{AD}$ is undefined {AD is vertical}

Let D be the mid-point of BC

Then D is the point (4, 2)

And eqn of AD is $\underline{x = 4}$ —— Eqn ①

$m_{AC} = \dfrac{9-2}{4-7} = \dfrac{6}{-3} = -2 \Rightarrow m_{BE} = \dfrac{1}{2}$

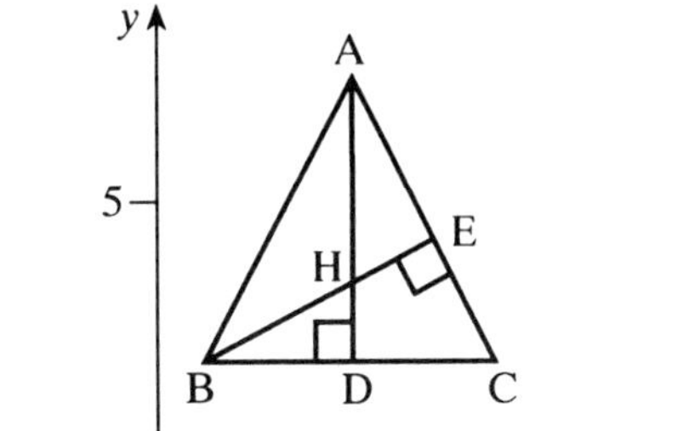


$$\text{Eqn}_{BE}\begin{cases} B(1, 2) & \\ & y - 2 = \dfrac{1}{2}(x-1) \\ m = \dfrac{1}{2} & 2y - 4 = x - 1 \end{cases}$$

$$\underline{2y = x + 3} \text{ —— Eqn ②}$$

By substitution from Eqn ① $\quad 2y = 4 + 3$

$$y = \underline{\frac{7}{2}}$$

Hence H is the point $\underline{\underline{\left(4, \; \dfrac{7}{2}\right)}}$

(ii) $\left.\begin{array}{l} A(4, 8);\ D(4, 2) \Rightarrow DA = 6 \\ H\left(4, \dfrac{7}{2}\right);\ D(4, 2) \Rightarrow DH = 1\dfrac{1}{2} \end{array}\right\} \Rightarrow$ $\underline{\underline{\text{H lies one quarter of the way up DA}}}$

**Question 4**

$$y = \frac{4}{x^2} + x\sqrt{x} = 4x^{-2} + x^{\frac{3}{2}}$$

$$\Rightarrow \frac{dy}{dx} = -8x^{-3} + \frac{3}{2}x^{\frac{1}{2}}$$

$$= \frac{-8}{x^3} + \frac{3}{2}\sqrt{x}$$

---

**Question 5**

*(a)*

$$f(x) = \frac{1}{x^2-4}; \; g(x) = 2x+1$$

$$h(x) = g(f(x)) = 2\left(\frac{1}{x^2-4}\right)+1$$

$$= \frac{2}{x^2-4}+1 = \frac{2}{x^2-4}+\frac{x^2-4}{x^2-4} = \frac{x^2-2}{x^2-4}$$

*(b)* We require $x^2 - 4 \neq 0$ for a suitable domain

Hence $x \neq -2$ or $2$

---

**Question 6**

$$\tan\alpha = \frac{\sqrt{11}}{3}$$

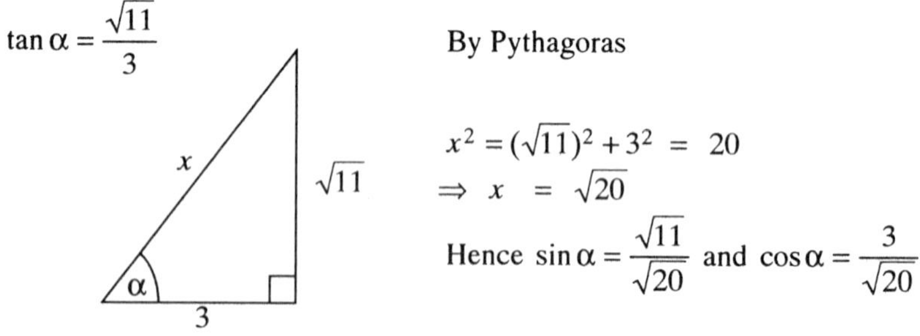


By Pythagoras

$$x^2 = (\sqrt{11})^2 + 3^2 = 20$$

$$\Rightarrow x = \sqrt{20}$$

Hence $\sin\alpha = \frac{\sqrt{11}}{\sqrt{20}}$ and $\cos\alpha = \frac{3}{\sqrt{20}}$

$$\sin 2\alpha = 2\sin\alpha\cos\alpha = 2\frac{\sqrt{11}}{\sqrt{20}} \cdot \frac{3}{\sqrt{20}} = \frac{6\sqrt{11}}{20}$$

$$\Rightarrow \sin 2\alpha = \frac{3\sqrt{11}}{10}$$

**Question 7**

$$m = f'(x)$$

$$m = \frac{2-0}{0-6} = \frac{2}{-6} = \frac{1}{-3}$$

$$\Rightarrow f'(x) = -\frac{1}{3}$$

---

**Question 8**

Since graph is a "straight line" $\{y = mx + c\}$

then $\log_{10} y = m \log_{10} x + c$

"gradient" $= 2$; so $m = 2$

"passes through" $(0,\ 1) \Rightarrow c = 1$

Let $c = \log_{10} A = 1$

$\Rightarrow A = 10$

$$\begin{aligned} \text{Hence} \quad \log_{10} y &= 2\log_{10} x + 1 && \{\log_{10} 10 = 1\} \\ &= 2\log_{10} x + \log_{10} 10 && \{2\log x = \log x^2\} \\ &= \log_{10} x^2 + \log_{10} 10 && \{\log a + \log b = \log ab\} \\ &= \log_{10} 10x^2 \\ \text{so} \quad \Rightarrow y &= \underline{\underline{10x^2}} \end{aligned}$$

---

**Question 9**

$f(x) = a\log_2(x - b)$ goes through $(3,\ 0)$, not $(1,\ 0)$
$\Rightarrow$ graph moved "$b$" units to the right, where $\underline{\underline{b = 2}}$

$$\begin{aligned} \text{At } (4,\ 3) \quad a\log_2(4-b) &= 3 \\ a\log_2(4-2) &= 3 \\ a\log_2 2 &= 3 && \{\log_2 2 = 1\} \\ a &= \underline{\underline{3}} \end{aligned}$$

---

**Question 10**

$$\begin{aligned}(x+1)(x+k) &= -4\\ x^2+x+kx+k+4 &= 0\\ x^2+(k+1)x+k+4 &= 0 \qquad \{\text{In the form } ax^2+bx+c=0\}\end{aligned}$$

Here $a=1$; $b=k+1$; $c=4+k$; for equal roots $b^2-4ac=0$

Hence
$$\begin{aligned}(k+1)^2-4(1)(4+k) &= 0\\ k^2+2k+1-16-4k &= 0\\ k^2-2k-15 &= 0\\ (k-5)(k+3) &= 0\end{aligned}$$

$$k = \underline{\underline{+5}} \quad \text{or} \quad k = \underline{\underline{-3}}$$

---

**Queston 11**

*(a)* Let

$$\begin{aligned}h(t) &= 20t-5t^2\\ h'(t) &= 20-10t \qquad \{\text{Thrown at } t=0\}\\ \Rightarrow h'(0) &= 20 \Rightarrow \text{initial speed} = \underline{\underline{20\text{ m s}^{-1}}}\end{aligned}$$

*(b)* From *(a)*

$$\begin{aligned}h'(t) &= 20-10t\\ h'(2) &= 20-10\times 2 = 0\end{aligned}$$

$\Rightarrow$ speed = $\underline{\underline{\ 0\ }}$ after 2 seconds. The ball is stationary.

{Maximum height is reached 20 m above thrown height}

---

**Question 12**

$$\begin{aligned}k\sin x^\circ &= 2 \Rightarrow k^2\sin^2 x^\circ = 4 \quad —— ③\\ k\cos x^\circ &= 2 \Rightarrow k^2\cos^2 x^\circ = 4 \quad —— ④\end{aligned}$$

$$\Rightarrow \tan x^\circ = \frac{2}{2}$$

$$x = \underline{\underline{45^\circ}}$$

③ + ④
$$\begin{aligned}k^2\sin^2 x^\circ + k^2\cos^2 x^\circ &= 8\\ k^2(\sin^2 x^\circ + \cos^2 x^\circ) &= 8\\ k^2 &= 8\\ k &= \underline{\underline{\sqrt{8}}}\end{aligned}$$

**Question 1**

$A_1(-4,\ -1);\ A_2(-1,\ 1);\ S(5,\ 4);$

$m_{A_1A_2} = \frac{1+1}{-1+4} = \frac{2}{3};\ m_{A_2S} = \frac{4-1}{5+1} = \frac{3}{6} = \frac{1}{2}$

Since $m_{A_1A_2} \neq m_{A_2S}$; then $A_1$, $A_2$ and S are NOT collinear

And "Achilles" will NOT pass directly over the submarine

$B_1(-7,\ -11);\ B_2(1,\ -1);\ S(5,\ 4);$

$m_{B_1B_2} = \frac{-1+11}{1+7} = \frac{10}{8} = \frac{5}{4};\ m_{B_2S} = \frac{4+1}{5-1} = \frac{5}{4}$

Since $m_{B_1B_2} = m_{B_2S}$; then $B_1$, $B_2$ and S are collinear

And "Belligerent" will pass directly over the submarine

---

**Question 2**

*(a)*

$$y = f(x) = x^3 - 2x^2 + x \Rightarrow f(2) = 8-8+2 = 2 \Rightarrow P(2,\ 2)$$

$$f'(x) = 3x^2 - 4x + 1 \Rightarrow f'(2) = 12-8+1 = 5 \Rightarrow m = 5$$

$$\text{Eqn}_{\text{tangent}} \begin{cases} P(2,\ 2) \\ m = 5 \end{cases} \quad y - 2 = 5(x-2)$$

$$y - 2 = 5x - 10$$

$$y = \underline{\underline{5x - 8}}$$

*(b)* They meet where :

$$x^3 - 2x^2 + x = 5x - 8$$

$$x^3 - 2x^2 - 4x + 8 = 0$$

$$(x-2)(x-2)(x+2) = 0$$

$$x = 2 \text{ (twice) or } x = -2$$

$$f(-2) = -8-8-2 = -18$$

Hence they meet again at $\underline{\underline{(-2,\ -18)}}$

Factorise

| | | | | |
|---|---|---|---|---|
| 2 | 1 | –2 | –4 | 8 |
| | | 2 | 0 | –8 |
| 2 | 1 | 0 | –4 | 0 |
| | | 2 | 4 | |
| | 1 | 2 | 0 | |

$(x+2)$

**Question 3**

For "Killpest"

65% destroyed $\Rightarrow$ 35% remain

Hence $K_{n+1} = 0\cdot 35K_n + 500$

In the long term :

$K_{n+1} \to K_n \to K$ is a limit value

$$\begin{aligned} \text{Hence} \quad K &= 0\cdot 35K + 500 \\ 0\cdot 65K &= 500 \\ K &= 769\cdot 2 \\ &\doteqdot \underline{\underline{769}} \end{aligned}$$

For "Pestkill"

85% destroyed $\Rightarrow$ 15% remain

Hence $P_{n+1} = 0\cdot 15P_n + 650$

In the long term :

$P_{n+1} \to P_n \to P$ is a limit value

$$\begin{aligned} \text{Hence} \quad P &= 0\cdot 15P + 650 \\ 0\cdot 85P &= 650 \\ P &= 764\cdot 7 \\ &\doteqdot \underline{\underline{765}} \end{aligned}$$

Hence "Pestkill" is marginally more effective in the long term

**Question 4**

*(a)* (i) Amplitude $= 2 \Rightarrow b = 2$, Frequency $= 3 \Rightarrow \underline{\underline{a = 3}}$ $\qquad f(x) = 2\sin 3x^\circ$

(ii) Amplitude $= 3 \Rightarrow d = 3$, Frequency $= 3 \Rightarrow \underline{\underline{c = 3}}$ $\qquad g(x) = 3\cos 3x^\circ$

*(b)*

$$\begin{aligned} h(x) &= f(x) + g(x) \\ &= 2\sin 3x^\circ + 3\cos 3x^\circ = q\sin(px + r)^\circ \end{aligned}$$

Hence $2\sin 3x^\circ + 3\cos 3x^\circ = q\sin px^\circ \cos r^\circ + q\cos px^\circ \sin r^\circ$

Here $\underline{\underline{p = 3}}$

$$\left.\begin{aligned} q\sin r^\circ &= 3 \\ q\cos r^\circ &= 2 \end{aligned}\right\} \Rightarrow \frac{q\sin r^\circ}{q\cos r^\circ} = \frac{3}{2}$$

$$\tan r^\circ = \frac{3}{2}$$

$$r^\circ = \underline{\underline{56\cdot 3}}$$

$$\left.\begin{aligned} q^2\sin^2 r^\circ &= 9 \\ q^2\cos^2 r^\circ &= 4 \end{aligned}\right\} \Rightarrow q^2(\sin^2 r + \cos^2 r) = 13$$

$$q = \underline{\underline{\sqrt{13}}}$$

Hence $\underline{\underline{h(x) = \sqrt{13}\sin(3x + 56\cdot 3)^\circ}}$

**Question 5**

*(a)*

$$|\vec{BR}|^2 = (7-5)^2 + (2-(-5))^2 + (3-(-1))^2 = 4 + 49 + 16 = 69$$

$$|\vec{BR}| = \sqrt{69} \Rightarrow \text{distance} = \underline{\underline{2\sqrt{69}\text{ km}}}$$

*(b)*

$$|\vec{MR}|^2 = (7-(-2))^2 + (2-4)^2 + (3\cdot 85)^2 = 81 + 4 + 30\cdot 25 = 115\cdot 25$$

$$|\vec{MR}| = \sqrt{115\cdot 25} \Rightarrow \text{distance} = 2\sqrt{115\cdot 25}\text{ km}$$

$$\text{Speed} = \frac{\text{distance}}{\text{time}} = \frac{2\sqrt{115\cdot 25}}{\frac{3}{60}} = 429\cdot 42$$

$$\doteqdot \underline{\underline{429\ \text{km/hr}}}$$

*(c)*

$$\vec{TC} = \underline{c} - \underline{t} = \begin{pmatrix}12\\-4\\1\end{pmatrix} - \begin{pmatrix}0\\0\\0\end{pmatrix} = \begin{pmatrix}12\\-4\\1\end{pmatrix}$$

$$\vec{BR} = \underline{r} - \underline{b} = \begin{pmatrix}7\\2\\3\end{pmatrix} - \begin{pmatrix}5\\-5\\-1\end{pmatrix} = \begin{pmatrix}2\\7\\4\end{pmatrix}$$

$$\vec{TC}\ .\ \vec{BR} = \begin{pmatrix}12\\-4\\1\end{pmatrix} . \begin{pmatrix}2\\7\\4\end{pmatrix} = 12\times 2 + (-4)\times 7 + 1\times 4 = 0$$

$\Rightarrow$ $\underline{\underline{\text{TC is perpendicular to BR}}}$ (since $\vec{TC}\ .\ \vec{BR} = 0$)

*(d)*

Let $\angle TCR = \theta°$; then $\cos\theta° = \dfrac{\vec{CT}\ .\ \vec{CR}}{|\vec{CT}|\ |\vec{CR}|}$ $\quad \left\{\vec{CR} = \begin{pmatrix}-5\\6\\2\end{pmatrix}\right\}$

$$= \frac{\begin{pmatrix}-12\\4\\-1\end{pmatrix} . \begin{pmatrix}-5\\6\\2\end{pmatrix}}{\sqrt{161}\ \ \sqrt{65}} \qquad \left\{\frac{82}{102\cdot 3}\right\}$$

$$= 0\cdot 802$$

$$\Rightarrow \theta = \underline{36\cdot 7}$$

Hence $\underline{\underline{\angle TCR = 36\cdot 7°}}$

**Question 6**

$$y = f(x) = ax^2 + bx + c \Rightarrow f(0) = \underline{\underline{c = 3}}$$

$$f'(x) = 2ax + b \qquad \begin{aligned} f'(-1) &= m_1 = \tan 45^\circ = 1 \\ f'(2) &= m_2 = \tan 135^\circ = -1 \\ \{f'(x) &= \text{gradient} = \tan\theta^\circ\} \end{aligned}$$

So
$$\begin{aligned} f'(-1) &= -2a + b = 1 \quad \text{——①} \\ f'(2) &= \underline{4a + b = -1} \quad \text{——②} \end{aligned}$$

$$\text{①} - \text{②} \qquad -6a = 2 \qquad\qquad \text{Substitute in ① — } -2\left(\frac{-1}{3}\right) + b = 1$$

$$\underline{a = -\frac{1}{3}} \qquad\qquad \underline{b = \frac{1}{3}}$$

Hence $f(x) = -\frac{1}{3}x^2 + \frac{1}{3}x + 3$ and $\underline{\underline{a = -\frac{1}{3};\ b = \frac{1}{3};\ c = 3}}$

**Question 7**

$$\text{Eqn } (x+12)^2+(y+15)^2 = 25 \Rightarrow A(-12,\ -15) \quad \underline{\underline{\text{Radius A} = 5}}$$
$$\text{Eqn } (x-24)^2+(y-12)^2 = 100 \Rightarrow C(24,\ 12) \quad \underline{\underline{\text{Radius C} = 10}}$$
$$A(-12,\ -15),\ C(24,\ 12) \Rightarrow AC^2 = 36^2+27^2 = 2025$$
$$\Rightarrow AC = \underline{\underline{45}}$$

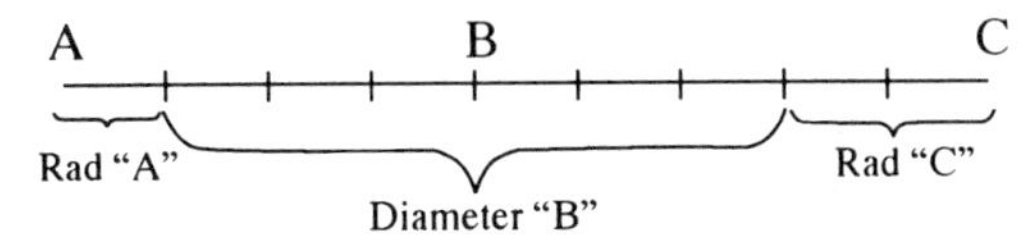


$$AC = \text{Rad "A"} + \text{Diam "B"} + \text{Rad "C"} = 45$$
$$5 + \text{Diam "B"} + 10 = 45$$
$$\Rightarrow \text{Diam "B"} = 30$$
$$\Rightarrow \text{Rad "B"} = \underline{\underline{15}}$$

Hence AB : BC

20 : 25

$\underline{\underline{4 : 5}}$

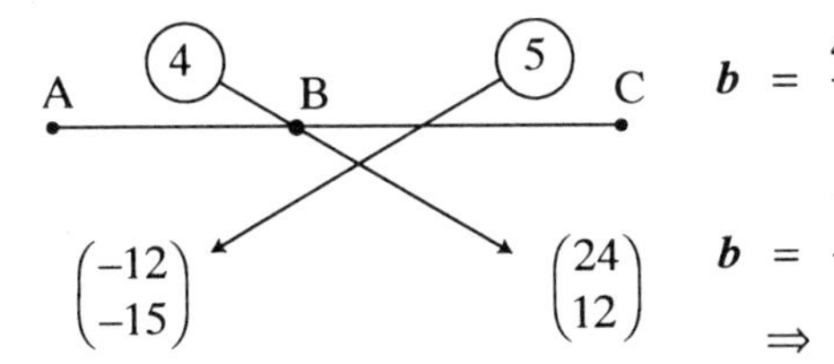


$$\boldsymbol{b} = \frac{4\boldsymbol{c}+5\boldsymbol{a}}{9}$$

$$\boldsymbol{b} = \frac{4\begin{pmatrix}24\\12\end{pmatrix}+5\begin{pmatrix}-12\\-15\end{pmatrix}}{9} = \frac{\begin{pmatrix}36\\-27\end{pmatrix}}{9} = \begin{pmatrix}4\\-3\end{pmatrix}$$

$\Rightarrow$ $\underline{\underline{\text{B is the point } (4,\ -3)}}$

Circle centre B(4, $-3$) and Radius = 15

Equation is $\underline{\underline{(x-4)^2+(y+3)^2 = 225}}$ $\{\text{or } x^2+y^2-8x+6y-200=0\}$

**Question 8**

*(a)* Area $= \int (y_{\text{curve}} - y_{\text{line}})\,dx$

The line meets the curve where $\left\{ y = 4 - \frac{1}{3}x = 0 \Rightarrow x = 12 \right\}$

$$4 - \frac{1}{3}x = 4 + \frac{5}{3}x - \frac{1}{6}x^2$$

$$\frac{1}{6}x^2 - \frac{6}{3}x = 0$$

$$x^2 - 12x = 0$$

$$x(x-12) = 0$$

$\underline{x = 0}$ or $\underline{x = 12}$

{These are the limits}

$$\text{Area} = \int_0^{12} \left(4 + \frac{5}{3}x - \frac{1}{6}x^2 - \left(4 - \frac{1}{3}x\right)\right)dx$$

$$= \int_0^{12} \left(2x - \frac{1}{6}x^2\right)dx$$

$$= \left[x^2 - \frac{x^3}{18}\right]_0^{12}$$

$$= \{(144 - 96) - (0)\}$$

$$\text{Area} = \underline{\underline{48 \text{ units}^2}}$$

*(b)* $\frac{1}{2}\text{Area} = \int_0^p \left(2x - \frac{x^2}{6}\right)dx = 24$

$$\Rightarrow p^2 - \frac{p^3}{18} = 24 \qquad \{\text{from } (a)\}$$

$$18p^2 - p^3 = 432$$

$$\underline{\underline{p^3 - 18p^2 + 432 = 0}}$$

*(c)* (i) Let $f(p) = p^3 - 18p^2 + 432$

| 6 | 1 | −18 | 0 | 432 |
|---|---|---|---|---|
| | | 6 | −72 | −432 |
| | 1 | −12 | −72 | 0 |

$\Rightarrow f(6) = 0$ i.e. $\underline{\underline{p = 6 \text{ is a solution}}}$

(ii) Let $p^2 - 12p - 72 = 0$ {Use the quadratic formula}

$\left.\begin{array}{l} a = 1 \\ b = -12 \\ c = -72 \end{array}\right\}$ $b^2 - 4ac = 432$ $\Rightarrow p = \frac{12 \pm \sqrt{432}}{2} = \frac{12 \pm 20 \cdot 78}{2}$

$$\Rightarrow p \doteq \underline{-4 \cdot 4 \text{ or } 16 \cdot 4}$$

(iii) Since $\underline{\underline{0 < p < 12;\ p = 6 \text{ is the only valid solution}}}$

$\{p = -4 \cdot 4 < 0;\ p = 16 \cdot 4 > 12\}$

**Question 9**

*(a)*

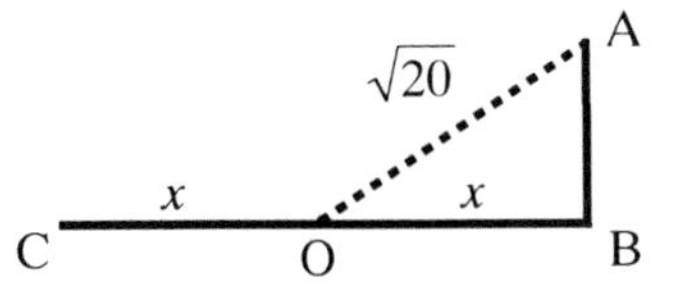


By Pythagoras

$$(\sqrt{20})^2 = AB^2 + x^2 = 20$$
$$\Rightarrow AB^2 = 20 - x^2$$
$$AB = \underline{\underline{\sqrt{20 - x^2}}}$$

Total length
$$T = CB + AB$$
$$= \underline{\underline{2x + \sqrt{20 - x^2}}}$$

*(b)*
$$T(x) = 2x + (20 - x^2)^{\frac{1}{2}}$$
$$T'(x) = 2 + \frac{1}{2}(20 - x^2)^{\frac{-1}{2}}(-2x)$$
$$\Rightarrow 2 - \frac{x}{\sqrt{20 - x^2}} = 0 \text{ at S.V.}$$
$$2\sqrt{20 - x^2} - x = 0$$
$$2\sqrt{20 - x^2} = x \quad \text{or} \quad \underline{\underline{x = 2\sqrt{20 - x^2}}}$$

*(c)* From *(b)*
$$x = 2\sqrt{20 - x^2}$$
$$x^2 = 4(20 - x^2)$$
$$x^2 = 80 - 4x^2$$
$$5x^2 = 80$$
$$x^2 = 16$$
$$x = \underline{\underline{4}}$$

{Ignore $x = -4$}

From *(a)*
$$T = 2x + \sqrt{20 - x^2}$$
at $x = 4$;
$$T = 2(4) + \sqrt{20 - (4)^2}$$
$$= 8 + \sqrt{4}$$
$$\underline{\underline{T = 10}}$$

$$\left\{\begin{array}{l} AB = \sqrt{20 - x^2} \\ AB = \sqrt{20 - 4^2} \end{array}\right\}$$
$$\left\{\begin{array}{l} AB = \sqrt{20 - 16} \\ AB = \sqrt{4} \\ AB = \underline{\underline{2}} \end{array}\right\}$$

| $x$ | $4^-$ | 4 | $4^+$ |
|---|---|---|---|
| $T'(x)$ | + | 0 | – |
| Shape | ↗ | → | ↘ |
| | | Max | |

$\underline{\underline{T = 10 \text{ is the max length}}}$ {A is the point (4, 2)}

# Model Paper D — Paper 1

(Non-calculator paper)

**Question 1**

Eqn of circle is $x^2 + y^2 + 2x - 4y - 15 = 0$

$\Rightarrow$ centre C(–1, 2) {centre $(-g,\ -f)$}

Since A(3, 4) and C(–1, 2)

$$m_{\text{radius AC}} = \frac{4-2}{3+1} = \frac{1}{2}$$

$\Rightarrow m_{\text{tangent}} = -2$ {since $m_1 \times m_2 = -1$}

$$\text{Eqn}_{\text{tangent}} \begin{cases} A(3,\ 4) \\ \\ m = -2 \end{cases} \qquad \begin{aligned} y-4 &= -2(x-3) \\ y-4 &= -2x+6 \\ \\ \underline{\underline{2x+y}} &\underline{\underline{\;= 10}} \end{aligned}$$

**Question 2**

$$\overset{\text{A}}{\boldsymbol{a}=\begin{pmatrix}2\\-5\\6\end{pmatrix}};\ \overset{\text{B}}{\boldsymbol{b}=\begin{pmatrix}6\\-3\\4\end{pmatrix}};\ \overset{\text{C}}{\boldsymbol{c}=\begin{pmatrix}12\\0\\1\end{pmatrix}}$$

$$\left.\begin{aligned} \overrightarrow{AB} &= \boldsymbol{b}-\boldsymbol{a} = \begin{pmatrix}6\\-3\\4\end{pmatrix}-\begin{pmatrix}2\\-5\\6\end{pmatrix} = \begin{pmatrix}4\\2\\-2\end{pmatrix} \\ \overrightarrow{BC} &= \boldsymbol{c}-\boldsymbol{b} = \begin{pmatrix}12\\0\\1\end{pmatrix}-\begin{pmatrix}6\\-3\\4\end{pmatrix} = \begin{pmatrix}6\\3\\-3\end{pmatrix} \end{aligned}\right\} \Rightarrow \underline{\underline{\overrightarrow{AB} = \frac{2}{3}\overrightarrow{BC}}}$$

Since $\overrightarrow{AB} = \frac{2}{3}\overrightarrow{BC}$

Then A, B and C are collinear

And B divides AC in the ratio $\underline{\underline{2:3}}$

**Question 3**

$$\begin{aligned} x^4 - x &= x(x^3-1) \\ &= \underline{\underline{x(x-1)(x^2+x+1)}} \end{aligned}$$

$$\begin{array}{r|rrrr} & x^3 & & & -1 \\ 1 & 1 & 0 & 0 & -1 \\ & & 1 & 1 & 1 \\ \hline & 1 & 1 & 1 & \underline{\underline{0}} \end{array}$$

**Question 4**

(i)

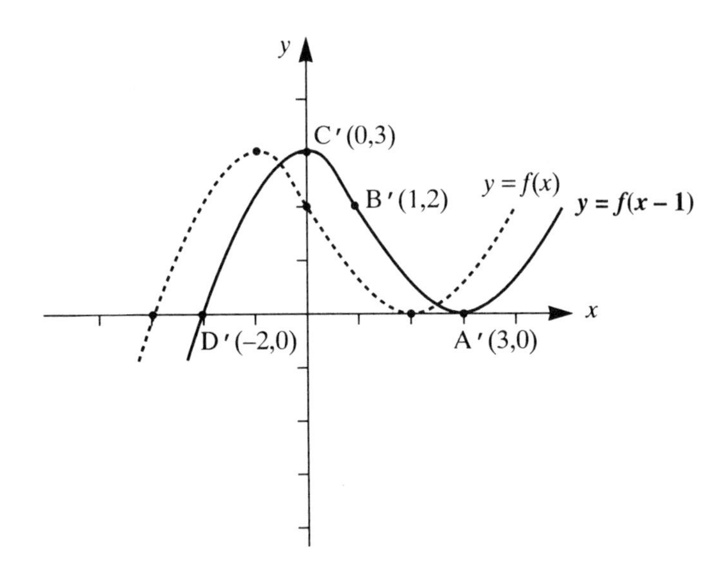


(ii)

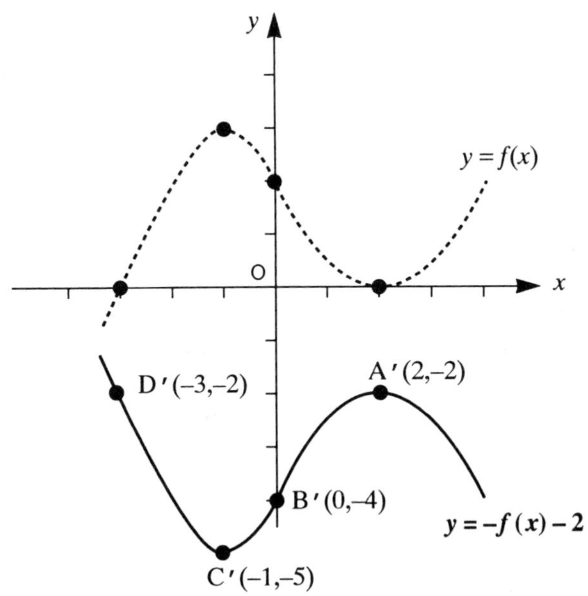

**Question 5**

$$f(x) = \frac{x-1}{\sqrt{x}} = \frac{x-1}{x^{\frac{1}{2}}} = x^{\frac{1}{2}} - x^{\frac{-1}{2}}$$

$\left\{\begin{array}{ll} \text{A} & \text{Algebra} \\ \text{B} & \text{Before} \\ \text{C} & \text{Calculus} \end{array}\right\}$

$$\begin{aligned} \text{Hence } f'(x) &= \frac{1}{2}x^{\frac{-1}{2}} + \frac{1}{2}x^{\frac{-3}{2}} \\ &= \frac{1}{2x^{\frac{1}{2}}} + \frac{1}{2x^{\frac{3}{2}}} \\ &= \underline{\frac{1}{2\sqrt{x}} + \frac{1}{2x\sqrt{x}}} \end{aligned}$$

$$\begin{aligned} \text{Hence } f'(4) &= \frac{1}{2\sqrt{4}} + \frac{1}{2(4)\sqrt{4}} \\ &= \frac{1}{4} + \frac{1}{16} \\ \Rightarrow f'(4) &= \underline{\underline{\frac{5}{16}}} \end{aligned}$$

---

**Question 6**

*(a)* This sequence has a limit as $n \to \infty$

since $\underline{0 \cdot 3 < 1}$ i.e. the coefficient of $u_n < 1$
$\{$sequence is linear in the form $u_{n+1} = mu_n + C\}$

*(b)* At this limit $u_{n+1} = u_n = u$ say (a fixed point)

$$\begin{aligned} \text{Hence } u &= 0 \cdot 3u + 5 \\ 0 \cdot 7u &= 5 \\ u &= \frac{5}{0 \cdot 7} \\ \text{So the limit} &= \underline{\underline{\frac{50}{7}}} \end{aligned}$$

$\left\{\frac{5}{0 \cdot 7} = \frac{50}{7}\right\}$

**Question 7**

*(a)* (i) $f(x) = 2x+1;\; g(x) = x^2+k$

$$g(f(x)) = (f(x))^2+k = (2x+1)^2+k = \underline{\underline{4x^2+4x+1+k}}$$

(ii) $$f(g(x)) = 2g(x)+1 = 2(x^2+k)+1 = \underline{\underline{2x^2+2k+1}}$$

*(b)* (i)
$$g(f(x))-f(g(x)) = 0$$
$$\Rightarrow 4x^2+4x+1+k-(2x^2+2k+1) = 0$$
$$\underline{\underline{2x^2+4x-k = 0}}$$

(ii) From (i) $2x^2+4x-k = 0$

$\left.\begin{array}{l} a=2 \\ b=4 \\ c=-6 \end{array}\right\}$ $2x^2+4x-6 = 0$

Here

$$b-4ac = 16-4(2)(-6) = 16+48 = \underline{\underline{64}} \text{ a perfect square}$$

Hence the roots are $\underline{\underline{\text{Real, Distinct and Rational}}}$

(iii) $2x^2+4x-k=0$

$\left.\begin{array}{l} a=2 \\ b=4 \\ c=-k \end{array}\right\}$ $b^2-4ac = 16-4(2)(-k) = 0$ for equal roots

$$16+8k = 0$$
$$8k = -16$$
$$k = \underline{\underline{-2}}$$

**Question 8**

In $\Delta BCD$; by Pythagoras

$$BD^2 = 3^2 + 4^2 = 25 \Rightarrow BD = \underline{\underline{5}}$$

In $\Delta ABD$; by Pythagoras

$$7^2 = AD^2 + 5^2 = 7^2$$
$$AD^2 + 25 = 49$$
$$AD^2 = 24$$
$$AD = \sqrt{24}$$
$$\Rightarrow AD = \underline{\underline{2\sqrt{6}}}$$

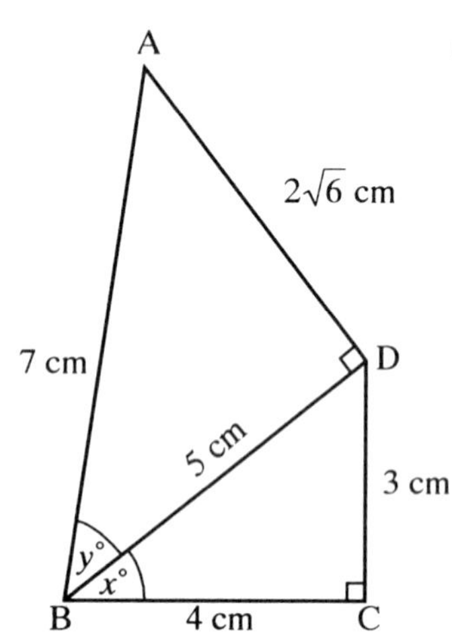


$$\cos(x+y)° = \cos x° \cos y° - \sin x° \sin y°$$
$$= \frac{4}{5} \cdot \frac{5}{7} - \frac{3}{5} \cdot \frac{2\sqrt{6}}{7}$$
$$= \frac{20}{35} - \frac{6\sqrt{6}}{35}$$
$$= \underline{\underline{\frac{20 - 6\sqrt{6}}{35}}}$$

**Question 9**

$$f(x) = 2x^3 - 3x^2 - 36x$$
$$\Rightarrow f'(x) = 6x^2 - 6x - 36 > 0 \text{ for } \underline{\underline{f(x) \text{ increasing}}}$$

$$6(x^2 - x - 6) > 0$$
$$6(x+2)(x-3) > 0$$
$$\underline{\underline{x < -2}} \quad \text{and} \quad \underline{\underline{x > 3}}$$

{A sketch will help}

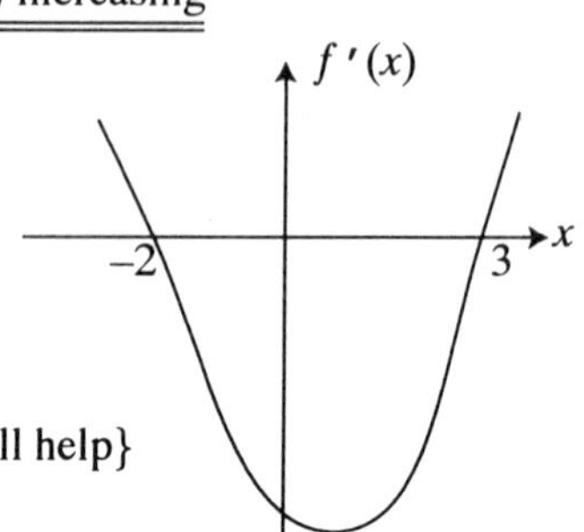

**Question 10**

Since $\Delta ABC$ is isosceles; then if D is the mid point of AC

$\Delta ABD$ is right angled at D

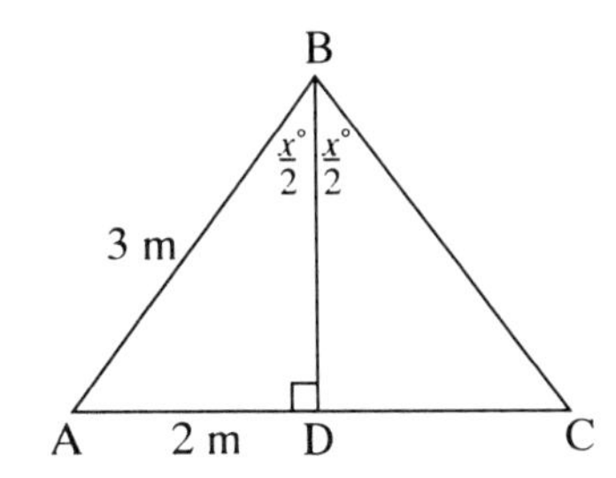


By Pythagoras; $3^2 = BD^2 + 2^2 = 3^2$

$$BD^2 + 4 = 9$$

$$BD^2 = 5$$

$$BD = \underline{\underline{\sqrt{5}}}$$

Hence $\sin x^\circ = \sin 2\left(\frac{x^\circ}{2}\right) = 2\sin\frac{x^\circ}{2}\cos\frac{x^\circ}{2}$

$$= 2 \cdot \frac{2}{3} \cdot \frac{\sqrt{5}}{3}$$

$$\Rightarrow \quad \sin x^\circ = \underline{\underline{\frac{4\sqrt{5}}{9}}}$$

{Alternatively use the Cosine Rule in $\Delta ABC$ to find $\cos x^\circ$; then $\sin^2 x^\circ = 1 - \cos^2 x^\circ$}

## Question 11

*(a)*

$$
\begin{aligned}
T_t &= T_0 10^{-kt} \Rightarrow 100 \times 10^{-10k} = 10\\
10^{-10k} &= \frac{1}{10}\\
10^{-10k} &= 10^{-1}\\
10k &= 1\\
k &= \underline{\underline{\frac{1}{10}}} \text{ or } 0\cdot 1
\end{aligned}
$$

*(b)*

$$
\begin{aligned}
T_t &= T_0 10^{-\frac{1}{10}t}\\
\Rightarrow T_{10} &= 10 \times 10^{-\frac{1}{10}\times 10}\\
&= 10 \times 10^{-1}\\
T_{10} &= 1 \Rightarrow \text{Fall in temperature is } \underline{\underline{9\ ^\circ\text{C}}} \qquad \{10-1\}
\end{aligned}
$$

## Question 12

Let M be the mid-point of the chord from O(0, 0) to A(6, 0)
then M is the point (3, 0)
Let the centre of the circle be C($x$, $y$)

then $x = 3$ {CM is $\perp$ bisector}

and the radius is $\underline{y+1}$
{$y = -1$ is a tangent}

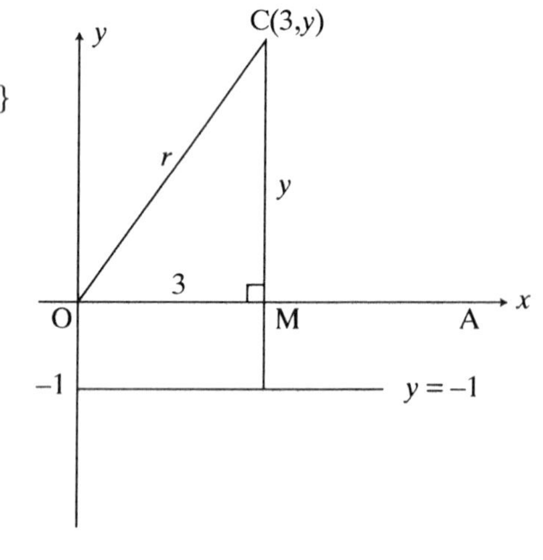


By Pythagoras

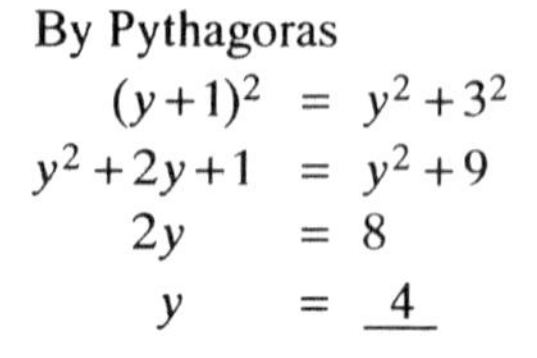

$$
\begin{aligned}
(y+1)^2 &= y^2 + 3^2\\
y^2 + 2y + 1 &= y^2 + 9\\
2y &= 8\\
y &= \underline{4}
\end{aligned}
$$

Hence centre (3, 4); radius = 5
$\Rightarrow$ Eqn is $\underline{\underline{(x-3)^2 + (y-4)^2 = 25}}$ {or $x^2 + y^2 - 6x - 8y = 0$}

**Question 1**

*(a)*

$$\begin{aligned} y = f(x) &= x^4 - 4x^3 + 3 \\ \Rightarrow f'(x) &= 4x^3 - 12x^2 = 0 \text{ at a Stationary Value} \\ & 4x^2(x-3) = 0 \\ & 4x^2 = 0 \quad \text{or} \quad x - 3 = 0 \\ & \underline{x = 0} \text{ (twice)} \quad \underline{x = 3} \\ & \underline{f(0) = 3} \quad \underline{f(3) = -24} \end{aligned}$$

Hence the points are $(0, 3)$ and $(3, -24)$

*(b)* For their nature use the nature table

| $x$ | $0^-$ | 0 | $0^+$ | $^-3$ | 3 | $3^+$ |
|---|---|---|---|---|---|---|
| $f'(x)$ | – | 0 | – | – | 0 | + |
| shape | ↘ | → | ↘ | ↘ | → | ↗ |
| type | Pt of Inf. | | | Min. T.P. | | |
| | (0, 3) | | | (3, –24) | | |

{Or use $f''(x) = 12x(x-2)$
$f''(0) = 0; f''(3) > 0$ etc.}

---

**Question 2**

*(a)*

$$A(-3, -3);\ B(-1, 1) \Rightarrow m_{AB} = \frac{1+3}{-1+3} = \frac{4}{2} = 2$$

$$C(7, -3);\ B(-1, 1) \Rightarrow m_{CB} = \frac{1+3}{-1-7} = \frac{4}{-8} = -\frac{1}{2}$$

Since $m_{AB} \times m_{BC} = 2 \times \left(-\frac{1}{2}\right) = -1$

Then $\Delta$ABC is right angled at B

*(b)* (i) B(–1, 1); C(7, –3) $\Rightarrow$ D(3, –1) {D is the mid-point of BC}

D(3, –1); A(–3 – 3) $\Rightarrow m_{AD} = \dfrac{-1+3}{3+3} = \dfrac{1}{3}$

$$\text{Eqn}_{AD}\begin{cases} A(-3,\ -3) \\ m = \dfrac{1}{3} \end{cases} \qquad \begin{aligned} y+3 &= \frac{1}{3}(x+3) \\ 3y+9 &= x+3 \end{aligned}$$

$$\underline{\underline{3y = x-6}} \text{ —— Eqn ①}$$

A(–3, –3); C(7, –3) $\Rightarrow$ E(2, –3) {E is the mid-point of AC}

E(2, –3); B(–1, 1) $\Rightarrow m_{BE} = \dfrac{1+3}{-1-2} = \dfrac{-4}{3}$

$$\text{Eqn}_{BE}\begin{cases} B(-1,\ 1) \\ m = \dfrac{-4}{3} \end{cases} \qquad \begin{aligned} y-1 &= \frac{-4}{3}(x+1) \\ 3y-3 &= -4x-4 \end{aligned}$$

$$\underline{\underline{3y = -4x-1}} \text{ —— Eqn ②}$$

(or $4x+3y+1=0$)

(ii) From (i) they meet where:

$$\begin{aligned} 3y = x-6 &= -4x-1 \qquad \{\text{from ① and ②}\} \\ 5x &= 5 \\ x &= 1 \\ y &= \frac{-5}{3} \end{aligned} \Rightarrow \underline{\underline{M\left(1,\ \frac{-5}{3}\right)}}$$

---

**Question 3**

*(a)* Q is 2 along, 2 in and 3 steps up

$\Rightarrow$ $\underline{\underline{\text{Q(2, 2, 9)}}}$ and $\underline{\underline{\text{R(21, 3, 12)}}}$

{R is 21 along, 3 in and 4 steps up}

*(b)*

$$\vec{PQ} = \boldsymbol{q} - \boldsymbol{p} = \begin{pmatrix}2\\2\\9\end{pmatrix} - \begin{pmatrix}12\\0\\0\end{pmatrix} = \underline{\underline{\begin{pmatrix}-10\\2\\9\end{pmatrix}}}$$

$$|\vec{PQ}| = \sqrt{(-10)^2 + 2^2 + 9^2} = \underline{\underline{\sqrt{185}}}$$

$$\vec{PR} = \boldsymbol{r} - \boldsymbol{p} = \begin{pmatrix}21\\3\\12\end{pmatrix} - \begin{pmatrix}12\\0\\0\end{pmatrix} = \underline{\underline{\begin{pmatrix}9\\3\\12\end{pmatrix}}}$$

$$|\vec{PR}| = \sqrt{9^2 + 3^2 + 12^2} = \underline{\underline{\sqrt{234}}}$$

$$\vec{PQ} \cdot \vec{PR} = \begin{pmatrix}-10\\2\\9\end{pmatrix} \cdot \begin{pmatrix}9\\3\\12\end{pmatrix} = -90 + 6 + 108 = \underline{\underline{24}}$$

By the Scalar Product

$$\cos\theta^\circ = \frac{\vec{PQ} \cdot \vec{PR}}{|\vec{PQ}|\,|\vec{PR}|} = \frac{24}{\sqrt{185}\sqrt{234}} = 0\cdot11535$$

$$\Rightarrow \theta = \underline{\underline{83\cdot4}}$$

Hence $\hat{QPR} = \underline{\underline{83\cdot4^\circ}}$

---

**Question 4**

$$\text{Area} = \int_{\pi/6}^{\pi/4} \cos 2x\,dx - \int_{\pi/4}^{\pi/2} \cos 2x\,dx$$

$$= \left[\frac{\sin 2x}{2}\right]_{\pi/6}^{\pi/4} - \left[\frac{\sin 2x}{2}\right]_{\pi/4}^{\pi/2}$$

$$= \left\{\left(\frac{1}{2}\right) - \left(\frac{\sqrt{3}}{4}\right)\right\} - \left\{(0) - \left(\frac{1}{2}\right)\right\}$$

$$\Rightarrow \text{Area} = \underline{\underline{1 - \frac{\sqrt{3}}{4} \text{ units}^2}} \qquad (\doteqdot 0\cdot567 \text{ unit}^2)$$

## Question 5

*(a)* 40% removed ⇒ 60% remain

Let the level after $n$ weeks $= U_n$ mg / $l$ $\quad (U_0 = 0)$

After $n$ weeks $U_n = 0\cdot6\, U_{n+1} + 2\cdot5 \quad (U_4 = 5\cdot44)$

If a fixed value "$U$" is reached then $U = 0\cdot6U + 2\cdot5$

$$0\cdot4U = 2\cdot5$$

Limit $U = 6\cdot25 \Rightarrow$ danger

*(b)* 30% clean ⇒ 70% remain

70% of $2\cdot5 = 1\cdot75$

Now $U_n = 0\cdot6\, U_{n-1} + 1\cdot75$

If a fixed value "$U$" is reached then $U = 0\cdot6U + 1\cdot75$

$$0\cdot4U = 1\cdot75$$

Limit $U = 4\cdot375 \Rightarrow$ safe

Hence; YES the L.A. should grant permission.

## Question 6

*(a)*

$$3\sin 2x^\circ = 2\sin x^\circ$$

$$3\sin 2x - 2\sin x^\circ = 0 \qquad \{\sin 2x^\circ = 2\sin x^\circ \cos x^\circ\}$$

$$3.2\sin x^\circ \cos x^\circ - 2\sin x^\circ = 0$$

$$2\sin x^\circ(3\cos x^\circ - 1) = 0$$

$$2\sin x^\circ = 0 \quad \text{or} \quad 3\cos x^\circ - 1 = 0$$

$$\sin x^\circ = 0 \qquad \cos x^\circ = \frac{1}{3}$$

$$x = 0,\ 180,\ 360 \qquad x = 70\cdot5,\ 289\cdot5$$

Hence $x = 0,\ 70\cdot5,\ 180,\ 289\cdot5,\ 360$

*(b)* $f(x) = 2\sin x^\circ$ ; $g(x) = 3\sin 2x^\circ$

*(c)* From *(a)* A = (70·5, 1·89 and B = (289·5, –1·89) $\quad \{f(70\cdot5) = 1\cdot89$ etc. . . .$\}$

*(d)* $3\sin2x^\circ < 2\sin x^\circ$ where $g(x) < f(x)$

⇒ where $70\cdot5 < x < 180$ AND $289\cdot5 < x < 360$

**Question 7**

*(a)* $A = (0, -50); m_{PB} = m_{AB} = \frac{4}{3}$ $\left\{\vec{AB} = \begin{pmatrix} 3 \\ 4 \end{pmatrix}\right\}$

Eqn$_{PB}$ is $\underline{\underline{y = \frac{4}{3}x - 50}}$ $\{y = mx + c\}$

*(b)* Eqn$_{circle}$ is $\underline{\underline{x^2 + y^2 = 900}}$ $\{x^2 + y^2 = r^2\}$

*(c)* $y = \frac{4}{3}x - 50 \Rightarrow y^2 = \frac{16}{9}x^2 - \frac{400}{3}x + 2500$

$$x^2 + y^2 = 900 \Rightarrow x^2 + \frac{16}{9}x^2 - \frac{400}{3}x + 2500 = 900$$

$$9x^2 + 16x^2 - 1200x + 14\ 400 = 0$$

$$25x^2 - 1200x + 14\ 400 = 0$$

$$x^2 - 48x + 576 = 0$$

$$(x - 24)^2 = 0$$

$$x = \underline{24} \text{ (twice)}$$

Equal roots $\Rightarrow$ tangent at $x = 24$ where $y = -18$

Hence $P = \underline{\underline{(24, -18)}}$

**Question 8**

*(a)* $$\text{Area} = \int_0^2 (2x - x^2)\,dx = \left[x^2 - \frac{x^3}{3}\right]_0^2 = 4 - \frac{8}{3} = \underline{\underline{\frac{4}{3}}} \text{ units}^2$$

*(b)* $$\text{Area} = \frac{1}{2}p \times \frac{1}{2}p = \frac{4}{3}$$

$$\Rightarrow \quad \frac{p^2}{4} = \frac{4}{3}$$

$$p^2 = \frac{16}{3}$$

$$p = \frac{4}{\sqrt{3}}$$

$$p \doteq \underline{\underline{2 \cdot 3 \text{ units}}}$$

*(c)* Area = $\int_{\frac{\pi}{4}}^{q} (\sin x - \cos x)\, dx = \frac{4}{3}$ {same area as *(a)*}

$$\left[-\cos x - \sin x\right]_{\pi/4}^{q} = \frac{4}{3}$$

$$\Rightarrow \quad \left(-\cos q - \sin q\right) - \left(-\cos\frac{\pi}{4} - \sin\frac{\pi}{4}\right) = \frac{4}{3}$$

$$\Rightarrow - \quad \left(-\cos q - \sin q\right) + \left(\frac{1}{\sqrt{2}} + \frac{1}{\sqrt{2}}\right) = \frac{4}{3}$$

$$\Rightarrow \quad \cos q + \sin q = \frac{2}{\sqrt{2}} - \frac{4}{3} \qquad \left\{\frac{2}{\sqrt{2}} = \sqrt{2}\right\}$$

$$\Rightarrow \quad \underline{\cos q + \sin q = 0 \cdot 081} \qquad \{1 \cdot 414 - 1 \cdot 333\}$$

Let $\cos q + \sin q = R\cos(q - \alpha)$

$$R^2 = 1^2 + 1^2 = 2 \; ; \quad \tan\alpha = \frac{1}{1}$$

$$\Rightarrow \quad \underline{R = \sqrt{2}} \qquad \underline{\alpha = \frac{\pi}{4}}$$

Hence $\cos q + \sin q = \sqrt{2}\cos\left(q - \frac{\pi}{4}\right) = 0 \cdot 081$

$$\cos\left(q - \frac{\pi}{4}\right) = 0 \cdot 0573$$

$$q - \frac{\pi}{4} = 1 \cdot 5135 \qquad \{\text{radians}\}$$

$$q = 1 \cdot 5135 + \frac{\pi}{4}$$

$$q = 2 \cdot 299$$

$$\underline{\underline{q \doteq 2 \cdot 3}}$$

---

**Question 1**

*(a)* E is the mid-point of AB $\Rightarrow$ E = (2, –1)

$$m_{CE} = \frac{8-(-1)}{-1-2} = -3 \qquad \left\{\text{Use}: m = \frac{y_2 - y_1}{x_2 - x_1}\right\}$$

$$\begin{aligned} \text{Eqn}_{CE} \text{ is} \quad y-8 &= -3(x-(-1)) \qquad \{\text{Use}: y-b = m(x-a)\} \\ y-8 &= -3x-3 \\ y &= \underline{\underline{-3x+5}} \end{aligned}$$

$$m_{AC} = \frac{8-0}{-1-7} = -1 \qquad \{m_1 \times m_2 = -1\}$$

$$\Rightarrow \quad m_{BD} = 1$$

$$\begin{aligned} \text{Eqn}_{BD} \text{ is} \quad y-(-2) &= 1(x-(-3)) \qquad \{y-b = m(x-a)\} \\ y+2 &= x+3 \\ y &= \underline{\underline{x+1}} \end{aligned}$$

*(b)* BD meets CE at J where

$$\begin{aligned} x+1 &= -3x+5 \qquad \{\text{from } (a)\} \\ 4x &= 4 \\ x &= 1 \\ \Rightarrow \underline{y} &\underline{= 2} \qquad \{\text{by substitution}\} \end{aligned}$$

Hence J = $\underline{\underline{(1, 2)}}$

**Question 2**

*(a)*

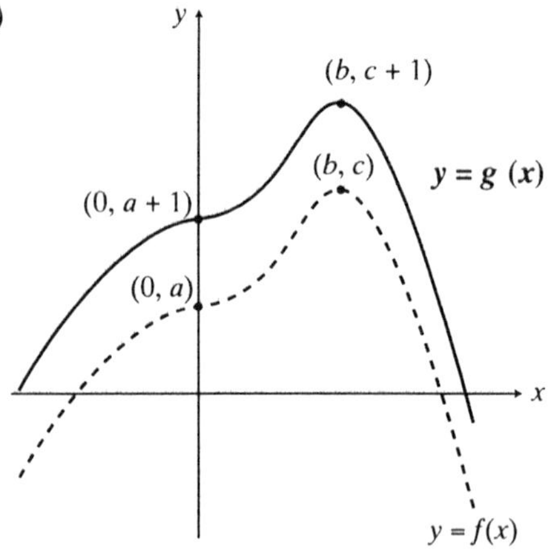


*(b)*

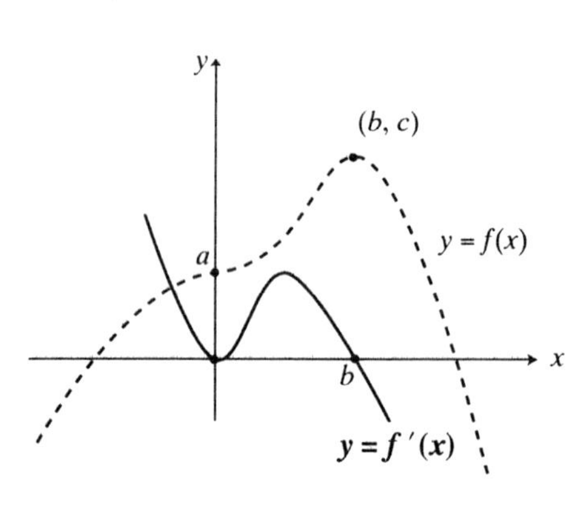


*(c)* They are the SAME, i.e. $f'(x) = g'(x)$

**Question 3**

$$\begin{aligned} \text{At A;}\quad y &= 5\log_{10}(2x+10) = 0 \\ \log_{10}(2x+10) &= 0 \\ 2x+10 &= 1 \\ 2x &= -9 \\ x &= -4\cdot 5 \\ &\Rightarrow \underline{\underline{A(-4\cdot 5,\ 0)}} \end{aligned}$$

$$\begin{aligned} \text{At B;}\quad y &= 5\log_{10}(2x+10) = 10 \\ \log_{10}(2x+10) &= 2 \\ 2x+10 &= 100 \\ 2x &= 90 \\ x &= 45 \\ &\Rightarrow \underline{\underline{B(45,\ 0)}} \end{aligned}$$

**Question 4**

For the line; $5x + y + 12 = 0$

$$y = -5x - 12$$

$$\Rightarrow \text{gradient} = \underline{-5}$$

For the curve; $y = 4x^2 + 3x - 5$

$$\Rightarrow \text{gradient} = \frac{dy}{dx} = \underline{8x + 3}$$

At the point of contact the gradients agree; $8x + 3 = -5$

$$8x = -8$$

$$\underline{\underline{x = -1}}$$

**Question 5**

*(a)* Let the perpendicular from L meet AB at P

Hence $\angle PBL = 180 - b° \Rightarrow \tan PBL = -\tan b°$

In $\Delta BLP$, $\tan B = \frac{d}{BP}$

$$\Rightarrow BP = \frac{d}{\tan B}$$

$$\Rightarrow BP = \frac{d}{\tan b°}$$

In $\Delta ALP$, $\tan A = \frac{d}{AP}$

$$\Rightarrow AP = \frac{d}{\tan a°}$$

$AB = AP + PB$

$$= \underline{\underline{\frac{d}{\tan a°} - \frac{d}{\tan b°}}}$$

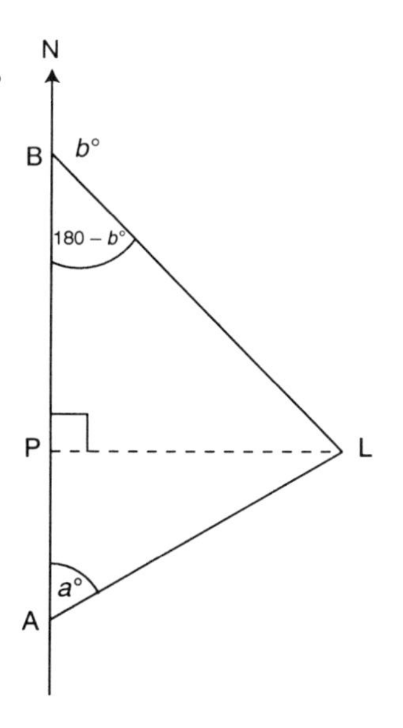

*(b)*

$$AB = \frac{d}{\tan a^\circ} - \frac{d}{\tan b^\circ} = \frac{d}{\frac{\sin a^\circ}{\cos a^\circ}} - \frac{d}{\frac{\sin b^\circ}{\cos b^\circ}} \qquad \left\{\tan a^\circ = \frac{\sin a^\circ}{\cos a^\circ}\right\}$$

$$= \frac{d\frac{\sin b^\circ}{\cos b^\circ} - d\frac{\sin a^\circ}{\cos a^\circ}}{\frac{\sin a^\circ}{\cos a^\circ}\frac{\sin b^\circ}{\cos b^\circ}} \qquad \left\{\times \frac{\cos a^\circ.\cos b^\circ}{\cos a^\circ \cos b^\circ}\right\}$$

$$= \frac{d \sin b^\circ \cos a^\circ - d \cos b^\circ \sin a^\circ}{\sin a^\circ \sin b^\circ}$$

$$= \underline{\underline{\frac{d \sin(b-a)^\circ}{\sin a^\circ \sin b^\circ}}}$$

---

**Question 6**

$$y = 2x^2 + x$$

$$\frac{dy}{dx} = \underline{\underline{4x+1}}$$

$$\text{LHS} = x\left(1 + \frac{dy}{dx}\right) = x(1+4x+1) = \underline{\underline{4x^2+2x}}$$

$$\text{RHS} = 2y = 2(2x^2+x) = \underline{\underline{4x^2+2x}} = \text{LHS}$$

---

**Question 7**

If $f'(x) = \cos 2x$; then $f(x) = \int \cos 2x \, dx$

Hence general solution is $f(x) = \frac{\sin 2x}{2} + c$

$$\text{At } \left(\frac{\pi}{12}, 1\right);\ f\left(\frac{\pi}{12}\right) = \frac{\sin 2\frac{\pi}{12}}{2} + c = 1$$

$$\Rightarrow \frac{1}{2}\cdot\frac{1}{2} + c = 1$$

$$\Rightarrow c = \frac{3}{4}$$

Hence particular solution is $f(x) = \underline{\underline{\frac{\sin 2x}{2} + \frac{3}{4}}}$

---

**Question 8**

If equation $C_1$ is $x^2 + y^2 - 10x - 4y + 12 = 0$

then centre is $C_1(5,\ 2)$ and radius $= \sqrt{5^2 + 2^2 - 12}$

$= \sqrt{17}$

P is the mid-point of the line of centres

$\sqrt{17}$ $\sqrt{17}$

$C_1$ (5, 2) P (9, 3) $C_2$ $(x, y)$

Hence $\dfrac{5+x}{2} = 9$ and $\dfrac{2+y}{2} = 3$

$5 + x = 18$ $2 + y = 6$

$x = 13$ $y = 4$

Hence $C_2$ is (13, 4)

Equation is $\underline{\underline{(x-13)^2 + (y-4)^2 = 17}}$

**Question 9**

$$\begin{aligned}\boldsymbol{a}.(\boldsymbol{b}+\boldsymbol{c}) &= \boldsymbol{a}.\boldsymbol{b} + \boldsymbol{a}.\boldsymbol{c} \\ &= 2\times 2\cos 60^\circ + 2\times 2\cos 120^\circ \\ &= 2 - 2 \\ &= 0\end{aligned}$$

Hence $\underline{\underline{\boldsymbol{a} \text{ is perpendicular to } (\boldsymbol{b}+\boldsymbol{c})}}$

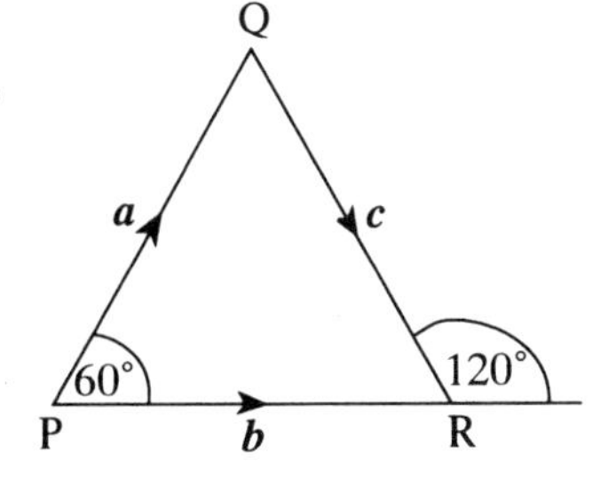


**Question 10**

$$\begin{aligned} f(x) &= 8 + 4\cos\frac{1}{2}x \\ \Rightarrow \quad f'(x) &= -2\sin\frac{1}{2}x \\ f'\left(\frac{7\pi}{3}\right) &= -2\sin\frac{7\pi}{6} = -2\left(-\frac{1}{2}\right) = 1 \\ f'(x) &= \text{gradient} = \tan a^\circ = 1 \\ a &= \underline{\underline{45^\circ}} \end{aligned}$$

**Question 11**

*(a)* $y = f(x) \quad = \frac{1}{x} = x^{-1}$

$\{m = f'(x)\} \quad \Rightarrow f'(x) = -x^{-2} = \frac{-1}{x^2};\ f'(a) = \underline{\underline{\frac{-1}{a^2}}}$

*(b)* $\text{Eqn}_{\text{tangent}}$ $\begin{cases} A\left(a,\ \frac{1}{a}\right) \\ m = \frac{-1}{a^2} \end{cases}$

$$y - \frac{1}{a} = \frac{-1}{a^2}(x-a)$$

$$a^2y - a = -x + a$$

$$\underline{\underline{x + a^2y = 2a}}$$

*(c)* (i) At B, $x = 0$ hence $a^2y = 2a$

$$y = \frac{2}{a} \Rightarrow B\left(0,\ \frac{2}{a}\right)$$

At C, $y = 0$ hence $x = 2a \Rightarrow C(2a,\ 0)$

Area of $\Delta$ $= \frac{1}{2}$ base $\times$ altitude

$= \frac{1}{2} \times 2a \times \frac{2}{a}$

Area of $\Delta$OBC $=$ $\underline{\underline{2 \text{ units}^2}}$ No "$a$"s in this answer

(ii) This area is <u>independent of the position of A</u> {always 2}

**Question 12**

Let $kx^2 - 8x + k = 0$

For NO REAL ROOTS $\quad 64 - 4k^2 < 0 \qquad \{b^2 - 4ac < 0\}$

$$4k^2 > 64$$

$$k^2 > 16$$

$$k > 4$$

Hence $\underline{\underline{k = 5}}$ $\qquad \{k > 0\}$

**Question 1**

*(a)* If $(x-2)$ is a factor, then $f(2)=0$

$$\begin{array}{c|cccl} 2 & 1 & k & -4 & -12 \\ & & 2 & 2k+4 & 4k \\ \hline & 1 & k+2 & 2k & 4k-12=0 \end{array}$$

{use: synthetic division}

$$4k = 12$$

$$\underline{\underline{k = 3}}$$

*(b)* If $k=3$ then $f(x) = x^3+3x^2-4x-12$

$= (x-2)(x^2+5x+6)$ {from *(a)*}

$= \underline{\underline{(x-2)(x+2)(x+3)}}$

**Question 2**

*(a)*

$$y = 2x+1 \Rightarrow -2y = -4x-2$$

also $y^2 = 4x^2+4x+1$

Equation $x^2+y^2+10x-2y-14 = 0$

becomes $x^2+4x^2+4x+1+10x-4x-2-14 = 0$

$$5x^2+10x-15 = 0$$

$$x^2+2x-3 = 0$$

$$(x-1)(x+3) = 0$$

$$x-1 = 0 \text{ or } x+3 = 0$$

$$x = 1 \text{ or } x = -3$$

$$y = 3 \text{ or } y = -5$$

Hence $\underline{\underline{A(1,\ 3)}}$ and $\underline{\underline{B(-3,\ -5)}}$

*(b)* (i) Centre $\underline{\underline{C(-5,\ 1)}}$

(ii) $m_{AB} = \dfrac{3-(-5)}{1-(-3)} = \dfrac{8}{4} = 2 \Rightarrow m_{l_2} = \underline{\underline{\dfrac{-1}{2}}}$ {Use $m_1 \times m_2 = -1$}

$\text{Eqn}_{l_2} \begin{cases} C(-5,\ 1) \\ m = \dfrac{-1}{2} \end{cases}$

$$y-1 = \frac{-1}{2}(x+5) \quad \{\text{Use } y-b = m(x-a)\}$$

$$2y-2 = -x-5$$

$$x+2y = -3 \text{ or } \underline{\underline{x+2y+3 = 0}}$$

**Question 3**

*(a)*

$$\vec{PR} = \frac{4}{3}\vec{PQ} \Rightarrow \boldsymbol{r} - \boldsymbol{p} = \frac{4}{3}(\boldsymbol{q} - \boldsymbol{p})$$

$$\boldsymbol{r} = \frac{4}{3}\left(\boldsymbol{q} - \frac{1}{3}\boldsymbol{p}\right)$$

$$\boldsymbol{r} = \frac{4}{3}\begin{pmatrix}5\\0\\5\end{pmatrix} - \frac{1}{3}\begin{pmatrix}-1\\3\\2\end{pmatrix}$$

$$\boldsymbol{r} = \begin{pmatrix}7\\-1\\6\end{pmatrix} \Rightarrow \underline{\underline{R(7,\ -1,\ 6)}}$$

*(b)*

$$\vec{SP} = \boldsymbol{p} - \boldsymbol{s} = \begin{pmatrix}-1\\3\\2\end{pmatrix} - \begin{pmatrix}-2\\2\\5\end{pmatrix} = \begin{pmatrix}1\\1\\-3\end{pmatrix};\ |\vec{SP}| = \sqrt{1+1+9} = \sqrt{11}$$

$$\vec{SR} = \boldsymbol{r} - \boldsymbol{s} = \begin{pmatrix}7\\-1\\6\end{pmatrix} - \begin{pmatrix}-2\\2\\5\end{pmatrix} = \begin{pmatrix}9\\-3\\1\end{pmatrix};\ |\vec{SR}| = \sqrt{81+9+1} = \sqrt{91}$$

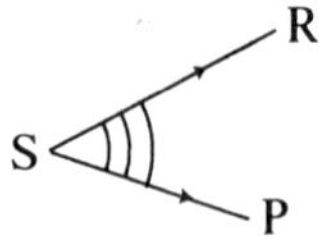

$$\vec{SP}.\vec{SR} = 1\times 9 + 1\times(-3) + (-3)\times 1 = 3$$

$$\text{Cos}\ P\hat{S}R^\circ = \frac{\vec{SP}.\vec{SR}}{|\vec{SP}||\vec{SR}|} = \frac{3}{\sqrt{11}\times\sqrt{91}} = 0\cdot 095$$

Hence $\underline{\underline{P\hat{S}R = 84\cdot 6^\circ}}$

**Question 4**

*(a)*

| | £ first day | | £ last day interest | | |
|---|---|---|---|---|---|
| Jan | 1000 | | 5 | | {Find 5% then add £100} |
| Feb | 1105 | | 5·525 | | |
| Mar | 1210·525 | | 6·052625 | | |
| Apr | 1316·577625 | | 6·5829 | | |
| May | 1423·1605 | | 7·116 | | |
| June | 1530·276 | + | 7·65 | = | £1537·93 |

*(b)*

| | | | |
|---|---|---|---|
| Jul | 1637·9 | | 8·18965 |
| Aug | 1746·12 | | 8·7306 |
| Sept | 1854·85 | | 9·2743 |
| Oct | 1964·12 | | 9·8206 |
| Nov | 2073·94 | $> 2000 \Rightarrow$ | November 1st |

*(c)* $A_{n+1} = 1{\cdot}005\,A_n + 100$ ; $A_0 = 1000$

$A_{n+1}$: Amount (£) in on 1st day of month

$A_n$: Amount (£) in on 1st day of previous month

$A_0$: Original amount (£)

---

**Question 5**

*(a)*

$$\begin{aligned}\sin x^\circ - 3\cos x^\circ &= k\sin(x-\alpha)^\circ \\ &= k\sin x^\circ\cos\alpha^\circ - k\cos x^\circ\sin\alpha^\circ\end{aligned}$$

$$\Rightarrow k\sin\alpha^\circ = 3$$

$$k\cos\alpha^\circ = 1$$

Hence (i) $k^2 = 3^2 + 1^2 = 10$ $\quad\{\sin^2\alpha^\circ + \cos^2\alpha^\circ = 1\}$

$$\Rightarrow k = \underline{\underline{\sqrt{10}}}$$

(ii) $\tan\alpha^\circ = 3$ $\quad\left\{\dfrac{\sin\alpha^\circ}{\cos\alpha^\circ} = \tan\alpha^\circ\right\}$

$$\Rightarrow \alpha = \underline{\underline{71\cdot6}}$$

Hence $\underline{\underline{\sin x^\circ - 3\cos x^\circ = \sqrt{10}\sin(x-71\cdot6)^\circ}}$

*(b)* Max. val. of $5 + \sin x^\circ - 3\cos x^\circ$ = max. val of $5 + \sqrt{10}\sin(x - 71\cdot6)^\circ$

$$= \underline{\underline{5 + \sqrt{10}}}$$

Max. val occurs where $(x - 71\cdot6)^\circ = 90^\circ$

$$x = \underline{\underline{161\cdot6}}$$

---

**Question 6**

*(a)* $f(x) = (x-2)^2 + 1 \Rightarrow f(0) = 5$ Hence $\underline{\underline{A(0, 5)}}$

$\Rightarrow \underline{\underline{B(2, 1)}}$ {Min T.P. 2 along, 1 up from origin}

*(b)* The curves meet where:

$$\begin{aligned}
\{(x-2)^2 + 1 &= x^2 - 4x + 5\} \\
5 + 4x - x^2 - (x^2 - 4x + 5) &= 0 \\
8x - 2x^2 &= 0 \\
2x(4-x) &= 0 \\
x = 0 \quad \text{or} \quad x &= 4 \\
\{y = 5 \quad \text{or} \quad y &= 5\}
\end{aligned}$$

$$\text{Area} = \int_0^4 (g(x) - f(x))dx = \int_0^4 (8x - 2x^2)dx$$

$$= \left[4x^2 - \frac{2x^3}{3}\right]_0^4$$

$$\left[4x^2 - \frac{2x^3}{3}\right]_0^4 = \left\{\left(64 - \frac{2 \times 64}{3}\right) - (0)\right\} = \frac{64}{3} \quad \text{or} \quad \underline{\underline{21\tfrac{1}{3} \text{ units}^2}}$$

*(c)* If $m + n \times f(x) = g(x)$

then $m + n(x^2 - 4x + 5) = 5 - 4x - x^2$

$m + nx^2 - 4nx + 5n = 5 - 4x - x^2$

Equating coefficients $n = \underline{\underline{-1}}$ {Reflection in $x$-axis}

$m + 5n = 5 \Rightarrow m = \underline{\underline{10}}$ {Move 10 up}

**Question 7**

*(a)*

$$y = y_0e^{kt} = \frac{1}{2}y_0 \quad \text{when } t = 5700$$

$$\Rightarrow e^{5700k} = 0 \cdot 5$$

$$5700\,k = \ln(0 \cdot 5) \qquad \{\text{Take } \log_e \text{ of both sides; } \log_e(e) = 1\}$$

$$k = -1 \cdot 216 \times 10^{-4}$$

$$k \doteq \underline{\underline{-0 \cdot 000122}} \text{ to 3 sig. figs.}$$

*(b)*

Let $y_0 = 100$ and $t = 1000$

$$y = y_0e^{kt} \Rightarrow y = 100e^{-0 \cdot 000122 \times 1000}$$

$$= 100e^{-0 \cdot 122}$$

$$= 88 \cdot 5 \Rightarrow \underline{\underline{88\tfrac{1}{2}\%}} \text{ remains}$$

---

**Question 8**

*(a)*

$$p = \frac{(4-(-2))}{2}; \qquad \underline{\underline{q = 1}}; \qquad r = -140; \qquad u = 230$$

$$\Rightarrow \underline{\underline{p = 3}}$$

{moved up 1} {Min T.P. (–90°) moved 140° to the right} {Max T.P. (90°) moved 140° to the right}

*(b)*

$$y = f(x) = 3\sin(x-140)^\circ + 1$$

For $s$; $x = 0$; $f(0) = 3\sin(-140)^\circ + 1 = -0 \cdot 928 \Rightarrow s = \underline{\underline{-0 \cdot 928}}$

For $t$; $y = 0$; $f(x) = 3\sin(x-140)^\circ + 1 = 0$

$$3\sin(x-140)^\circ = -1$$

$$\sin(x-140)^\circ = \frac{-1}{3}$$

$$(x-140)^\circ = -19 \cdot 5$$

$$x = 120 \cdot 5 \Rightarrow t = \underline{\underline{120 \cdot 5}}$$

---

**Question 9**

*(a)* $$\int_0^{\frac{\pi}{2}} \cos 2x \ \ dx = \left[\frac{1}{2}\sin 2x\right]_0^{\frac{\pi}{2}} = \frac{1}{2}(\sin \pi) - \frac{1}{2}(\sin 0)$$

$$= \underline{\underline{0}}$$

*(b)*

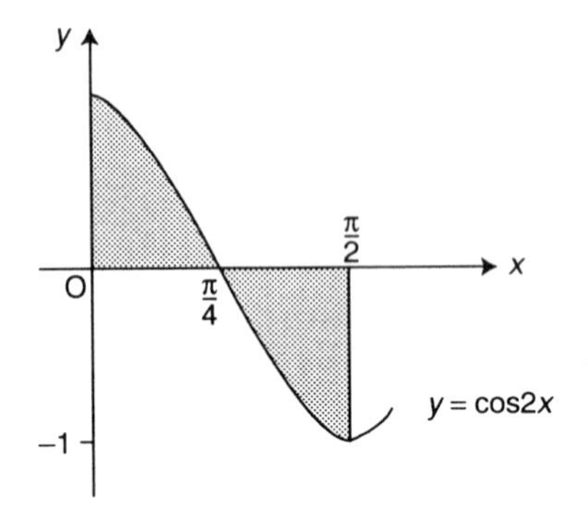


Positive and negative areas cancel.

**Question 10**

*(a)* (i) By similar $\Delta$s $\frac{10-h}{x} = \frac{10}{4} = \frac{5}{2}$

$$\Rightarrow 10-h = \frac{5}{2}x$$

$$h = 10 - \frac{5}{2}x$$

(ii)

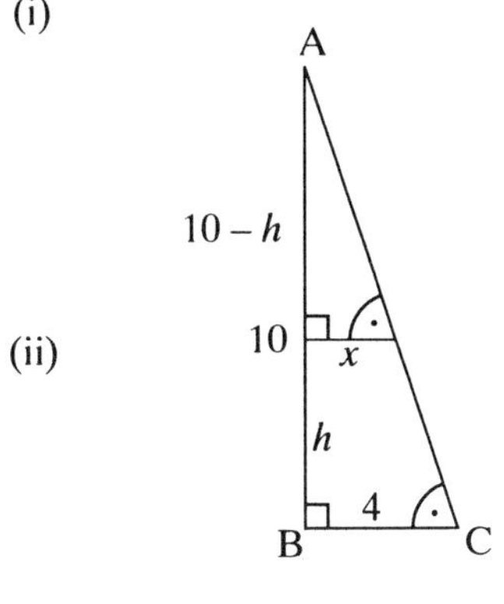


Volume = area of base × height

$$= (2x)^2 \times \left(10 - \frac{5}{2}x\right)$$

$$= 4x^2 \times \left(10 - \frac{5}{2}x\right)$$

$$\Rightarrow \text{vol} = 40x^2 - 10x^3$$

*(b)*

$$V(x) = 40x^2 - 10x^3$$

$$V'(x) = 80x - 30x^2 = 0 \text{ at St. Val.}$$

$$10x(8-3x) = 0$$

$$10x = 0 \text{ or } (8-3x) = 0$$

$$x = 0 \text{ or } x = \frac{8}{3}$$

Nature Table

| $x$ | $\frac{8}{3}^-$ | $\frac{8}{3}$ | $\frac{8}{3}^+$ |
|---|---|---|---|
| $V^1(x)$ | + | 0 | – |
| Shape | ↗ | → | ↘ |

Max T.P. at $x = \frac{8}{3}$

$$h = 10 - \frac{5}{2} \times \frac{8}{3} = \frac{10}{3}$$

Hence cuboid dimensions $2x \times 2x \times h$ in cm

$$= \frac{16}{3} \text{ cm} \times \frac{16}{3} \text{ cm} \times \frac{10}{3} \text{ cm}$$

## Model Paper F — Paper 1

(Non-calculator paper)

---

**Question 1**

Since AM is a median; then M is the mid-point of BC

$$\Rightarrow M\left(\frac{6\ -\ 2}{2}, \frac{1\ -\ 3}{2}\right)$$

$$\Rightarrow M(2, -1)$$

If A(4, 3) and M(2, –1); then gradient$_{AM}$ $= \dfrac{3-(-1)}{4-2} = 2$

Equation$_{AM}$ is
$$y-3 = 2(x-4) \quad \{\text{Use: } y-b = m(x-a)\}$$
$$y-3 = 2x-8$$
$$\underline{\underline{y = 2x-5}}$$

---

**Question 2**

$f(x) = x^3 - 4x^2 - 7x + 10$ {Try factors of 10, e.g., ±1, ±2}

$$\begin{array}{r|rrrr}
1 & 1 & -4 & -7 & 10 \\
 & & 1 & -3 & -10 \\
\hline
-2 & 1 & -3 & -10 & 0 \\
 & & -2 & 10 & \\
\hline
 & 1 & -5 & 0 &
\end{array}$$

$1\ \ -3\ \ -10\ \ 0 \Rightarrow (x-1)$ is a factor

$1\ \ -5\ \ 0 \Rightarrow (x+2)$ is a factor

$\Rightarrow (x-5)$ is a factor

Hence $\underline{\underline{f(x) = (x+2)(x-1)(x-5)}}$

---

**Question 3** If equation is $(x-3)^2 + (y+2)^2 = 25$; then centre is $\underline{C(3–2)}$

Let P be the point (6, 2); then $m_{PC} = \dfrac{2-(-2)}{6-3} = \dfrac{4}{3}$

If $m_{radius} = \dfrac{4}{3}$; then $m_{tangent} = \dfrac{-3}{4}$ {since $m_r \times m_t = -1$)

Equation$_{tangent}$ is $y-2 = \dfrac{-3}{4}(x-6)$ {use $y-b = m(x-a)$}

$$4y-8 = -3x+18$$
$$4y = -3x+26$$
$$\underline{\underline{3x+4y = 26}}$$

---

**Question 4**

*(a)* (i) $y = f(x) = x^3 - 9x; \Rightarrow f(-2) = (-2)^3 - 9(-2) = 10 \Rightarrow \underline{\underline{A(-2,\ 10)}}$

(ii) $f'(x) = 3x^2 - 9; \Rightarrow f'(-2) = 3(-2)^2 - 9 = 3; \Rightarrow m = 3$

$$\text{Eqn}_{AB}\begin{cases} A(-2,\ 10) \\ m = 3 \end{cases} \quad \begin{aligned} y - 10 &= 3(x+2) \\ y - 10 &= 3x + 6 \\ \underline{\underline{y}} &\underline{\underline{= 3x + 16}} \end{aligned}$$

*(b)* $y = 3x + 16$ meets $y = x^3 - 9x$ where

$$\begin{aligned} x^3 - 9x &= 3x + 16 \\ x^3 - 12x - 16 &= 0 \\ (x+2)(x+2)(x-4) &= 0 \end{aligned}$$

$x = -2$ twice or $\left.\begin{aligned} x &= 4 \\ y &= 28 \end{aligned}\right\} \underline{\underline{B(4,\ 28)}}$

{Factorise using double root at $x = -2$ leaving $x - 4$ as a factor}

**Question 5**

*(a)*
$$\begin{aligned} f(g(x)) = (g(x))^2 - 1 &= (x^2 + 2)^2 - 1 \\ &= x^4 + 4x^2 + 4 - 1 \\ &= \underline{\underline{x^4 + 4x^2 + 3}} \end{aligned}$$

*(b)* $x^4 + 4x^2 + 3 = \underline{\underline{(x^2 + 1)(x^2 + 3)}}$

{Substitute $y = x^2$; then $y^2 + 4y + 3 = (y + 1)(y + 3)$}

**Question 6**

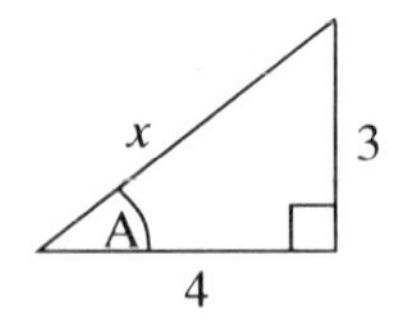


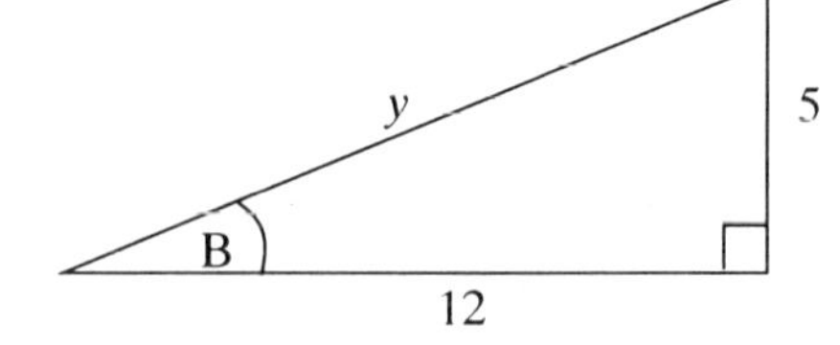


By Pythagoras $x^2 = 3^2 + 4^2 = 25; \quad y^2 = 12^2 + 5^2 = 169$

$\Rightarrow x = 5 \qquad \Rightarrow y = 13$

*(a)* $\quad \text{Sin } 2\text{A} = 2 \sin \text{A} \cos \text{A} = 2 \times \frac{3}{5} \times \frac{4}{5} = \frac{24}{25}$

*(b)* $\quad \text{Cos } 2\text{A} = \cos^2 \text{A} - \sin^2 \text{A} = \left(\frac{4}{5}\right)^2 - \left(\frac{3}{5}\right)^2 = \frac{7}{25}$

*(c)* $\quad \text{Sin } (2\text{A} + \text{B}) = \sin 2\text{A} \cos \text{B} + \cos 2\text{A} \sin \text{B}$

$$= \frac{24}{25} \times \frac{12}{13} + \frac{7}{25} \times \frac{5}{13}$$

$$= \frac{323}{325}$$

**Question 7**

*(a)* $\quad v_{n+1} = 0{\cdot}3\, v_n + 4$ approaches a limit since $0{\cdot}3 < 1$

{In the form $v_{n+1} = mv_n + c$; we require $-1 < m < 1;\ m \neq 0$}

*(b)* At the limit $v_{n+1} = v_n = v$

Hence $v = 0{\cdot}3v + 4$

$0{\cdot}7v = 4$

$$v = \frac{4}{0 \cdot 7} = \frac{40}{7} \text{ or } 5\frac{5}{7}$$

**Question 8**

$$\frac{dy}{dx} = 6x^2 - 2x$$

$$y = \int (6x^2 - 2x)dx$$

$$y = 2x^3 - x^2 + c$$

At $(-1, 2)$; $2 = 2(-1)^3 - (-1)^2 + c = 2$

$$-2 - 1 + c = 2$$

$$c = 5$$

Hence $\underline{\underline{y = 2x^3 - x^2 + 5}}$

**Question 9**

$$y = x^3 + kx^2 - 8x + 3$$ {$m$ of $x$-axis = 0; $m = 0$ at T.P.}

$$\frac{dy}{dx} = 3x^2 + 2kx - 8 = 0 \text{ at } x = -2$$

$$3(-2)^2 + 2k(-2) - 8 = 0$$

$$12 - 4k - 8 = 0$$

$$-4k + 4 = 0$$

$$-4k = -4$$

$$\underline{\underline{k = 1}}$$

**Question 10**

*(a)*

$$\begin{aligned} y = \sin 2x &= 0 \\ 2x &= 0, \pi, 2\pi \\ x &= 0, \frac{\pi}{2}, \pi \Rightarrow p = \frac{\pi}{2} \end{aligned}$$

$$\begin{aligned} y = \sin 2x &= -1 \\ 2x &= \frac{3\pi}{2}, \frac{7\pi}{2} \\ x &= \frac{3\pi}{4}, \frac{7\pi}{4} \Rightarrow \underline{\underline{q = \frac{3\pi}{4}}} \end{aligned}$$

$\left\{ \text{or use symmetry, } q \text{ is half-way between } \frac{\pi}{2} \text{ and } \pi \right\}$

*(b)*

$$\begin{aligned} \text{Area} = -\int_{\frac{\pi}{2}}^{\frac{3\pi}{4}} \sin 2x \, dx &= \frac{-1}{2}\left[-\cos 2x\right]_{\frac{\pi}{2}}^{\frac{3\pi}{4}} \\ &= \frac{-1}{2}\left\{\left(-\cos\frac{3\pi}{2}\right) - (-\cos\pi)\right\} \\ &= \frac{-1}{2}\{0-1\} \\ &= \underline{\underline{\frac{1}{2}\text{units}^2}} \end{aligned}$$

---

**Question 11**

$$\begin{aligned} \underset{\sim}{a}.(\underset{\sim}{a}+\underset{\sim}{b}) = \underset{\sim}{a}.\underset{\sim}{a}+\underset{\sim}{a}.\underset{\sim}{b} &= |\underset{\sim}{a}||\underset{\sim}{a}|\cos 0° + |\underset{\sim}{a}||\underset{\sim}{b}|\cos 60° \\ &= 2\times 2\times 1 \quad + \; 2\times 3\times\frac{1}{2} \\ &= \underline{\underline{7}} \end{aligned}$$

---

**Question 12**

*(a)*

$$\cos 2\theta + 8\cos\theta + 9 = 0$$
$$2\cos^2\theta - 1 + 8\cos\theta + 9 = 0$$
$$2\cos^2\theta + 8\cos\theta + 8 = 0$$
$$\cos^2\theta + 4\cos\theta + 4 = 0$$
$$(\cos\theta + 2)^2 = 0$$
$$\cos\theta = -2 \text{ twice}$$

{factorise or show $b^2 - 4ac = 0$}

$\Rightarrow$ equal roots since it is a perfect square.

*(b)* $\cos\theta = -2$

$\theta$ has NO real roots since $\cos\theta < -1$

**Question 1**

*(a)*

$$\overrightarrow{AB} = \boldsymbol{b} - \boldsymbol{a} = \begin{pmatrix}3\\6\\5\end{pmatrix} - \begin{pmatrix}2\\-1\\3\end{pmatrix} = \underline{\underline{\begin{pmatrix}1\\7\\2\end{pmatrix}}}$$

$$\overrightarrow{AC} = \boldsymbol{c} - \boldsymbol{a} = \begin{pmatrix}6\\6\\-2\end{pmatrix} - \begin{pmatrix}2\\-1\\3\end{pmatrix} = \underline{\underline{\begin{pmatrix}4\\7\\-5\end{pmatrix}}}$$

*(b)*

$$\cos\ B\hat{A}C = \frac{\overrightarrow{AB}\ .\ \overrightarrow{AC}}{|\overrightarrow{AB}||\overrightarrow{AC}|} = \frac{\begin{pmatrix}1\\7\\2\end{pmatrix} . \begin{pmatrix}4\\7\\-5\end{pmatrix}}{\left|\begin{pmatrix}1\\7\\2\end{pmatrix}\right|\left|\begin{pmatrix}4\\7\\-5\end{pmatrix}\right|}$$

$$= \frac{4 + 49 - 10}{\sqrt{54}\sqrt{90}} \qquad \{1^2 + 7^2 + 2^2 = 54 \text{ etc.}\}$$

$$= \frac{43}{\sqrt{4860}}$$

$$= 0 \cdot 617$$

$$\Rightarrow B\hat{A}C = \underline{\underline{51 \cdot 9^\circ}}$$

*(c)*

$$\text{Area } \Delta ABC = \frac{1}{2} bc \sin A^\circ$$

$$= \frac{1}{2}\sqrt{90}\ .\ \sqrt{54}\ \sin 51 \cdot 9^\circ$$

$$\Rightarrow \text{Area} = \underline{\underline{27 \cdot 4 \text{ units}^2}}$$

**Question 2**

*(a)*

$$\begin{aligned} y = f(x) &= -x^4 + 4x^3 - 2 \\ f'(x) &= -4x^3 + 12x^2 = 0 \text{ at St. Val.} \\ -4x^2(x-3) &= 0 \\ -4x^2 = 0 &\text{ or } x - 3 = 0 \\ x = 0 &\text{ or } x = 3 \\ y = -2 &\text{ or } y = 25 \end{aligned}$$

$\underline{\underline{\text{TPs } (0, -2) \text{ and } (3, 25)}}$

*(b)*

| $x$ | $0^-$ | 0 | $0^+$ |
|---|---|---|---|
| $f'(x)$ | + | 0 | + |
| Shape | ↗ | → | ↗ |

| $x$ | $3^-$ | 3 | $3^+$ |
|---|---|---|---|
| $f'(x)$ | + | 0 | – |
| Shape | ↗ | → | ↘ |

Point of Inflection at $\underline{\underline{(0, -2)}}$

Max TP at $\underline{\underline{(3, 25)}}$

**Question 3**

*(a)*

$$y = f(x) = x^3 - 4x^2 + 2x - 1; \quad \Rightarrow f(2) = 8 - 16 + 4 - 1 = -5 \Rightarrow P(2, -5)$$

$$f'(x) = 3x^2 - 8x + 2; \quad \Rightarrow f'(2) = 12 - 16 + 2 = -2 \Rightarrow m = -2$$

$$\text{Eqn}_{\text{tangent}} \begin{cases} P(2, \ -5) \\ & y - (-5) = -2(x-2) \\ & y + 5 = -2x + 4 \\ m = -2 & y = -2x - 1 \end{cases}$$

or $\underline{\underline{2x + y + 1 = 0}}$

*(b)*

$$m_{\text{tangent}} \times m_{\text{normal}} = -1 \Rightarrow m_{\text{normal}} = \frac{1}{2}$$

$$m = \tan\theta° = \frac{1}{2}$$

$$\theta = 26 \cdot 6$$

$$\Rightarrow \text{Required angle} = \underline{\underline{26 \cdot 6°}}$$

**Question 4**

*(a)*

$$y = f(x) = ax(b-x) \qquad \left\{\begin{array}{l} x=0 \text{ and } x=6 \text{ are roots} \\ \text{so } f(0)=f(6)=0 \end{array}\right\}$$

$$f(6) = 6a(b-6) = 0$$

$$6a = 0 \text{ or } b-6=0$$

$$a = 0 \text{ or } \underline{b=6} \qquad \{a \neq 0\}$$

$$f(3) = 3a(6-3) = 9$$

$$9a = 9$$

$$\underline{a = 1}$$

Hence $\underline{\underline{a=1 \text{ and } b=6}}$ and $f(x) = x(6-x)$
or $f(x) = 6x - x^2$

*(b)*

$$\text{Area} = \int_0^6 (6x-x^2)dx = \left[3x^2 - \frac{x^3}{3}\right]_0^6$$

$$= \left\{\left(3\times 36 - \frac{216}{3}\right) - (0)\right\}$$

$$= \underline{\underline{36 \text{ units}^2}}$$

*(c)* (i) $y = 6x - x^2$ meets $y = x$ where:

$$x = 6x - x^2$$

$$x^2 - 5x = 0$$

$$x(x-5) = 0$$

$$x = 0 \text{ or } x - 5 = 0$$

$$\left.\begin{array}{l} x=5 \\ y=5 \end{array}\right\} \Rightarrow \underline{\underline{\text{P}(5,\ 5)}}$$

(ii)

$$\text{Area} = \int_0^5 \{(6x-x^2)-(x)\ dx\} \qquad \left\{\text{Area} = \int (f(x)-g(x))dx\right\}$$

$$= \int_0^5 (5x-x^2)\,dx$$

$$= \left[\frac{5x^2}{2} - \frac{x^3}{3}\right]_0^5$$

$$= \frac{125}{2} - \frac{125}{3}$$

$$\Rightarrow \text{Area} = \underline{\underline{\frac{125}{6} \text{ units}^2}} \qquad \{20\cdot 83 \text{ units}^2\}$$

**Question 5**

*(a)* A is the mid-point of DE

$$\Rightarrow A\left(\frac{2-1}{2}, \frac{4+2}{2}\right) \text{ so } A\left(\frac{1}{2}, 3\right)$$

$$AE^2 = \left(2-\frac{1}{2}\right)^2 + (4-3)^2 \quad \{AE = \text{radius}\}$$

$$= \left(1\frac{1}{2}\right)^2 + 1^2$$

$$= 3\frac{1}{4} \text{ or } \frac{13}{4} \qquad \{\text{use } (x-a)^2 + (y-b)^2 = r^2\}$$

Hence $\text{Eqn}_{\text{circle}}$ is $\left(x-\frac{1}{2}\right)^2 + (y-3)^2 = \frac{13}{4}$

*(b)* (i) B(8, 8)

(ii) B is the mid-point of EF

Let F be $(x, y)$ $\Rightarrow \frac{2+x}{2} = 8$ and $\frac{4+y}{2} = 8$

$2 + x = 16 \qquad 4 + y = 16$

$x = 14 \qquad y = 12$

so F(14, 12)

C is the mid-point of DF

$$\Rightarrow C\left(\frac{-1+14}{2}, \frac{2+12}{2}\right) \text{ so } C\left(\frac{13}{2}, 7\right)$$

*(c)* If D(–1, 2) and F(14, 12) then $DF^2 = (14+1)^2 + (12-2)^2 = 325$

$$\Rightarrow DF = \sqrt{325} = \sqrt{25}\sqrt{13}$$

$$\Rightarrow DF = 5\sqrt{13}$$

$$\text{Perimeter} = \frac{1}{2}\{\pi DE + \pi EF + \pi DF\}$$

$$= \frac{\pi}{2}\{DE + EF + DF\} \qquad \left\{\begin{array}{l}\text{circumference} = \pi d \\ \text{semi-circle} = \frac{\pi}{2}d\end{array}\right\}$$

$$= \frac{\pi}{2}\{DF + DF\} \qquad \{DE + EF = DF\}$$

$$= \frac{\pi}{2}\{2DF\}$$

$$= \pi DF$$

$$= \underline{\underline{5\pi\sqrt{13}\text{ units}}} \quad \text{as required}$$

---

**Question 6**

*(a)* $f(x) = 2\cos x° - 3\sin x° = k\cos(x+\alpha)°$

Equating coefficients $= k\cos x°\cos\alpha° - k\sin x°\sin\alpha°$

Here $\left.\begin{array}{l}k\sin\alpha° = 3 \\ k\cos\alpha = 2\end{array}\right\}$ $\tan\alpha° = \frac{3}{2}$ Also $k^2 = 3^2 + 2^2 = 13$

$\Rightarrow \underline{\alpha = 56{\cdot}3}$ $\qquad \underline{\underline{k = \sqrt{13}}}$

Hence $f(x) = \underline{\underline{\sqrt{13}\cos(x + 56{\cdot}3)°}}$

*(b)* Max Val $= \sqrt{13}$ at $(x + 56{\cdot}3)° = 0°$ or $360°$

$x = \underline{\underline{-56{\cdot}3° \text{ or } 303{\cdot}7°}}$

Min Val $= -\sqrt{13}$ at $(x + 56{\cdot}3)° = 180°$

$x = \underline{\underline{123{\cdot}7°}}$

*(c)* Min Val of $(f(x))^2 = \underline{\underline{0}}$ $\qquad \left\{\begin{array}{l}\text{at } f(x) = 0 \\ \Rightarrow x = 33\cdot 7\end{array}\right\}$

---

**Question 7**

*(a)* Let the amount after the $n$th feeding be $An$

25% loss ⇒ 75% remains so if $A_1 = 1$

then $A_2 = 0{\cdot}75\,A_1 + 1 \;\Rightarrow\; A_2 = 1{\cdot}75$

$A_3 = 0{\cdot}75\,A_2 + 1 \;\Rightarrow\; A_3 = 2{\cdot}3125$

$A_4 = 0{\cdot}75\,A_3 + 1 \;\Rightarrow\; A_4 = 2{\cdot}734375$

After 4 feeds the amount stays > 2 g

$\{0{\cdot}75 \times A_4 = 2{\cdot}051\}$

So 4 feeds are necessary

*(b)* (i) $A_{n+1} = 0{\cdot}75\,A_n + 1$ {i.e. multiply by 0·75, then add 1}

(ii) Reaches a limit when $A_{n+1} = A_n = A$

$$A = 0{\cdot}75\,A + 1$$

$$0{\cdot}25A = 1$$

⇒ The limit is $A = 4$ Yes it is safe as limit < 5

$\{\text{check } 4 = 0{\cdot}75 \times 4 + 1\}$

**Question 8**

*(a)*

$$\text{Vol}_{\text{cylinder}} = \pi r^2 h = 400$$
$$\Rightarrow h = \frac{400}{\pi r^2}$$

$$\begin{aligned}\text{Surface area} &= \pi r^2 + 2\pi r h + 2\pi r^2 \\ &= 3\pi r^2 + 2\pi r h \\ &= 3\pi r^2 + 2\pi r \times \frac{400}{\pi r^2} \\ \Rightarrow A(r) &= \underline{\underline{3\pi r^2 + \frac{800}{r}}}\end{aligned}$$

*(b)*

$$\begin{aligned}A(r) &= 3\pi r^2 + 800 r^{-1} \\ A'(r) &= 6\pi r - 800 r^{-2} = 0 \text{ at St. Val.} \qquad \{\text{multiply by } r^2\} \\ 6\pi r^3 - 800 &= 0 \\ 6\pi r^3 &= 800 \\ r^3 &= \frac{800}{6\pi} \\ r &= \underline{\underline{3 \cdot 488}} \qquad (\doteqdot 3 \cdot 5 \text{ cm})\end{aligned}$$

| $x$ | $3{\cdot}5^-$ | $3{\cdot}5$ | $3{\cdot}5^+$ |
|---|---|---|---|
| $A'(x)$ | – | 0 | + |
| Shape | ↘ | → | ↗ |

Min turning value at $\underline{\underline{r \doteqdot 3 \cdot 5}}$

$\{$or use $A''(r) > 0 \Rightarrow$ min val$\}$

**Question 9**

*(a)*

$$\begin{aligned} y &= ae^{bx} \\ \log_e y &= \log_e(ae^{bx}) \\ \log_e y &= \log_e a + \log_e e^{bx} \\ \log_e y &= \log_e a + bx \end{aligned}$$

This is $\underline{\underline{\text{linear}}}$ in the form $y = mx + c$ where $m = b$ and $c = \log_e a$

*(b)* From the table

$x = 3{\cdot}1$      $x = 5{\cdot}2$

$y = 21\,876$      $y = 11\,913\,076$

$\log_e y = 9{\cdot}993$      $\log_e y = 16{\cdot}293$

Form 2 equations

$\log_e y = bx + \log_e a$

$$16{\cdot}293 = 5{\cdot}2b + \log_e a$$

$$9{\cdot}993 = 3{\cdot}1b + \log_e a$$

Subtract

$$6{\cdot}3 = 2{\cdot}1b$$

$$\underline{\underline{b = 3}}$$

By substitution

$$\begin{aligned} 9{\cdot}993 = 3{\cdot}1 \times 3 + \log_e a &= 9{\cdot}993 \\ \log_e a &= 0{\cdot}693 \\ \underline{\underline{a}} &\underline{\underline{= 2}} \end{aligned}$$

Hence $\underline{\underline{y = 2e^{3x}}}$

$$\left\{\begin{array}{l}\text{check if } x = 4{\cdot}1 \\ y = 2e^{12{\cdot}3} = 439\ 392\end{array}\right\}$$

## Model Paper G — Paper 1

(Non-calculator paper)

### Question 1

| | | $2x^3$ | $+\ x^2$ | $-\ 13x$ | $+\ 6$ | |
|---|---|---|---|---|---|---|
| *(a)* | 2 | 2 | 1 | –13 | 6 | |
| | | | 4 | 10 | –6 | |
| *(b)* | –3 | 2 | 5 | –3 | $\underline{\underline{0}}$ | $\Rightarrow (x-2)$ is a factor; $\underline{\underline{x = 2 \text{ is a root}}}$ |
| | | | –6 | 3 | | |
| | | 2 | –1 | 0 | | $\Rightarrow (x+3)$ is a factor; $\underline{\underline{x = -3 \text{ is a root}}}$ |

$\Rightarrow$ $2x-1$ is a factor

$\underline{\underline{x = \frac{1}{2} \text{ is a root}}}$

### Question 2

*(a)* $m_{AB} = \frac{5+1}{5+3} = \frac{6}{8} = \frac{3}{4}$

$$\text{Eqn}_{AB}\begin{cases} A(-3,-1) \\ m = \frac{3}{4} \end{cases} \qquad \begin{aligned} y+1 &= \frac{3}{4}(x+3) \\ 4y+4 &= 3x+9 \end{aligned}$$

$$\underline{\underline{4y = 3x+5}}$$

*(b)* Mid Pt. $M\left(\frac{5-3}{2}, \frac{5-1}{2}\right) \Rightarrow M(1,2)$

gradient will be $\frac{-4}{3}$ $\qquad$ $\{\text{since } M_1 \times M_2 = -1\}$

$$\text{Eqn}\begin{cases} M(1,2) \\ \frac{-4}{3} \end{cases} \qquad \begin{aligned} y-2 &= \frac{-4}{3}(x-1) \\ 3y-6 &= -4x+4 \end{aligned}$$

$$\underline{\underline{4x+3y = 10}}$$

**Question 3**

*(a)* $f(x) = 2x + 4$

$f(5) = 14$ so area $= \frac{1}{2}(14 + 8) \times 3$

$f(2) = 8$ $= \underline{\underline{33 \text{ units}^2}}$

*(b)* Area $= \underline{\underline{\int_2^5 (2x + 4)\,dx}}$

*(c)*
$$\int_2^5 (2x + 4)\,dx = [x^2 + 4x]_2^5$$
$$= (25 + 20) - (4 + 8)$$
$$= \underline{\underline{33}}$$

**Question 4**

*(a)* $x^2 + y^2 + 6x + 4y + 8 = 0 \Rightarrow$ centre $P(-3,\ -2)$

$\left.\begin{matrix} A(-1,\ -1) \\ P(-3,\ -2) \end{matrix}\right\}$ $m_{AP} = \dfrac{-1+2}{-1+3} = \dfrac{1}{2} \Rightarrow m_{\tan} = -2$ (since $m_1 \times m_2 = -1$)

$\text{Eqn}_{\tan}\begin{cases} A(-1,-1) \\ m = -2 \end{cases}$ $y + 1 = -2(x + 1)$

$y + 1 = -2x - 2$

$\underline{\underline{2x + y = -3}}$ OR $2x + y + 3 = 0$

*(b)* At B; $x = 0 \Rightarrow y = -3 \Rightarrow B(0, -3)$

*(c)* A is the mid point of BC (perp. to chord) $\Rightarrow C(-2, 1)$

*(d)* Circle centre $A(-1,-1)$ radius AB $= \sqrt{5}$

$\underline{\underline{(x+1)^2 + (y+1)^2 = 5}}$

**Question 5**

$$\begin{aligned}\vec{CV} &= \vec{CB} + \vec{BA} + \vec{AV} \\ &= \vec{DA} + \vec{BA} + \vec{AV} \\ &= -\vec{AD} - \vec{AB} + \vec{AV} \\ &= -\begin{pmatrix}-2\\10\\-2\end{pmatrix} - \begin{pmatrix}8\\2\\2\end{pmatrix} + \begin{pmatrix}1\\7\\7\end{pmatrix}\end{aligned}$$

$$\text{Hence } \vec{CV} = \begin{pmatrix}-5\\-5\\7\end{pmatrix}$$

**Question 6**

*(a)*

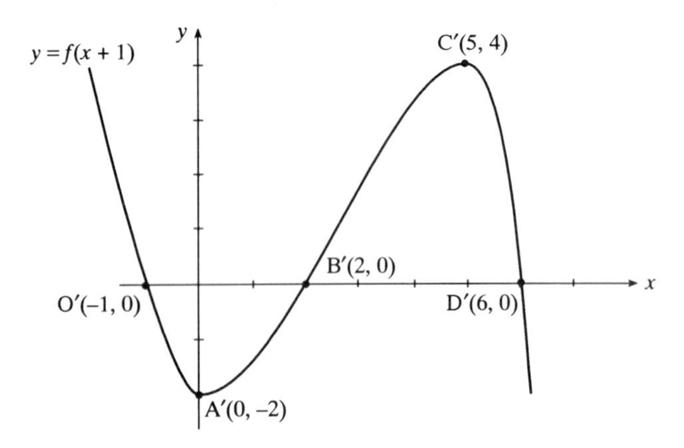

*(b)*

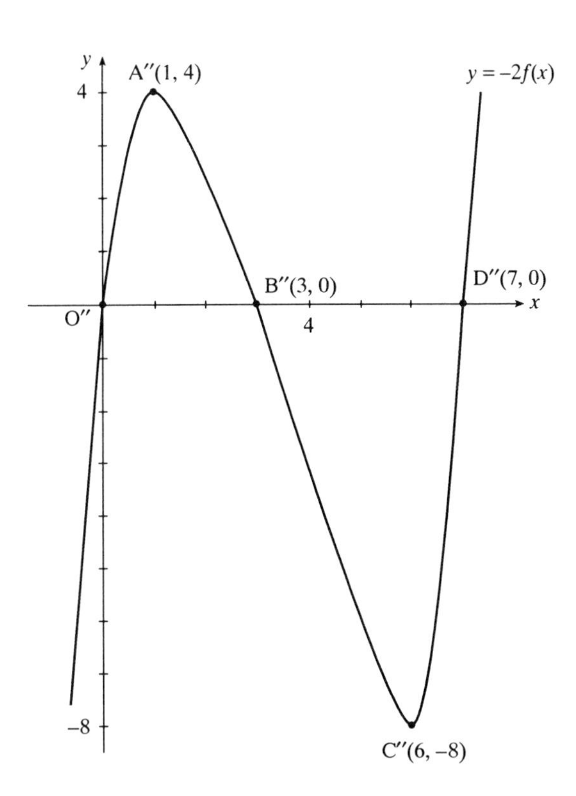


---

**Question 7**

$$y = \log_b(x + a)$$

At $x = 3$; $y = \log_b(3 + a) = 0 \Rightarrow 3 + a = 1 \Rightarrow a = -2$

At $x = 7$; $y = \log_b(7 - 2) = 1 \Rightarrow \log_b 5 = 1 \Rightarrow b = 5 \quad \{y = \log_5(x - 2)\}$

$$\underline{\underline{a = -2;\ b = 5}}$$

---

**Question 8**

*(a)* (i) $\underline{a} \cdot \underline{a} = |\underline{a}||\underline{a}|\cos 0 = a^2 = \underline{\underline{9}}$

(ii) $\underline{b} \cdot \underline{b} = |\underline{b}||\underline{b}|\cos 0 = b^2 = \underline{\underline{8}}$

(iii) $\underline{a} \cdot \underline{b} = |\underline{a}||\underline{b}|\cos 45° = 3 \times 2\sqrt{2} \cdot \frac{1}{\sqrt{2}} = \underline{\underline{6}}$

*(b)*

$$\underline{p} = 2\underline{a} + 3\underline{b}$$

$$\underline{p} \cdot \underline{p} = (2\underline{a} + 3\underline{b}) \cdot (2\underline{a} + 3\underline{b}) = 4\underline{a} \cdot \underline{a} + 12\underline{a} \cdot \underline{b} + 9\underline{b} \cdot \underline{b}$$

$$= 4(9) + 12(6) + 9(8)$$

$$\underline{p} \cdot \underline{p} = p^2 \qquad \Rightarrow \underline{p} \cdot \underline{p} = 180$$

$$\text{If } p^2 = 180$$

$$p = \sqrt{180} = \sqrt{36}\sqrt{5}$$

$$= \underline{\underline{6\sqrt{5}}}$$

**Question 9**

At a limit $u_{n+1} = u_n = u$ and $v_{n+1} - v_n = v$

$$u = 0 \cdot 2u + p \quad ; \quad v = 0 \cdot 6v + q$$

$$0 \cdot 8u = p \qquad 0 \cdot 4v = q$$

$$u = \frac{p}{0 \cdot 8} \qquad v = \frac{q}{0 \cdot 4}$$

But $u = v$

Hence $\frac{p}{0 \cdot 8} = \frac{q}{0 \cdot 4} \Rightarrow p = \frac{0 \cdot 8q}{0 \cdot 4} \Rightarrow p = \underline{\underline{2q}}$

**Question 10**

$$y = f(x) = 2x^3 + 3x^2 + 4x - 5$$
$$f'(x) = 6x^2 + 6x + 4 = 0 \text{ at St.V}$$

Here

$$\left.\begin{array}{l} a = 6 \\ b = 6 \\ c = 4 \end{array}\right\} \quad b^2 - 4ac = 36 - 4 \times 6 \times 4$$
$$= -12 < 0 \text{ so no real roots}$$
$$\Rightarrow \underline{\underline{\text{No stationary points}}}$$

**Question 11**

$$f(x) = \cos^2 x - \sin^2 x = \cos 2x$$
$$f'(x) = \underline{\underline{-2\sin 2x}}$$

## Question 1

*(a)* A(–4, 1) B(12, 3)

$\Rightarrow$ M(4, 2)

C(7, –7)

$$m_{CM} = \frac{2+7}{4-7} = \frac{9}{-3} = -3$$

$$\text{Eqn}_{CM}\begin{cases} M(4,\ 2) \\ m-3 \end{cases} \quad y-2 = -3(x-4)$$

$$y = -3x+14$$

$$\underline{\underline{3x+y = 14}} \quad \text{———} \quad ①$$

*(b)* $m_{BC} = \frac{3+7}{12-7} = \frac{10}{5} = 2 \Rightarrow m_{AD} = \frac{-1}{2}$ (sin θ $m_1 \times m_2 = -1$)

$$\text{Eqn}_{AD}\begin{cases} A(-4,1) \\ m = \frac{-1}{2} \end{cases} \quad y-1 = \frac{-1}{2}(x+4)$$

$$2y-2 = -x-4$$

$$\underline{\underline{x+2y = -2}} \quad \text{———} \quad ②$$

OR $$x+2y+2 = 0$$

*(c)* Where they meet

$$\begin{array}{lcrcl} 2① & \Rightarrow & 6x+2y & = & 28 \\ ② & \Rightarrow & x+2y & = & -2 \\ \hline \text{Subtract} & & 5x & = & 30 \\ & & x & = & 6 \\ & & y & = & -4 \end{array}$$

They meet at $\underline{\underline{(6, -4)}}$

## Question 2

$$y = f(x) = 5x^2+2$$

$$f'(x) = 10x \qquad f'(-1) = -10 = m$$

$$\begin{cases} P(-1,7) \\ m = -10 \end{cases} \quad \begin{aligned} y-7 &= -10(x+1) \\ y-7 &= -10x-10 \end{aligned}$$

$$\underline{\underline{10x+y+3=0}}$$

**Question 3**

*(a)*

$$\vec{AK} = \vec{AD} + \vec{DH} + \vec{HK}$$
$$= \vec{AD} + \vec{AE} + \frac{2}{3}\vec{HG}$$
$$= \vec{AD} + \vec{AE} + \frac{2}{3}\vec{AB}$$
$$= \begin{pmatrix} -8 \\ 4 \\ 4 \end{pmatrix} + \begin{pmatrix} 1 \\ -3 \\ 5 \end{pmatrix} + \frac{2}{3}\begin{pmatrix} 3 \\ 6 \\ 3 \end{pmatrix}$$
$$\vec{AK} = \underline{\underline{\begin{pmatrix} -5 \\ 5 \\ 11 \end{pmatrix}}}$$

*(b)*

$$\vec{AL} = \vec{AB} + \vec{BF} + \vec{FL}$$
$$= \vec{AB} + \vec{AE} + \frac{1}{4}\vec{FG}$$
$$= \vec{AB} + \vec{AE} + \frac{1}{4}\vec{AD}$$
$$= \begin{pmatrix} 3 \\ 6 \\ 3 \end{pmatrix} + \begin{pmatrix} 1 \\ -3 \\ 5 \end{pmatrix} + \frac{1}{4}\begin{pmatrix} -8 \\ 4 \\ 4 \end{pmatrix}$$
$$\vec{AL} = \underline{\underline{\begin{pmatrix} 2 \\ 4 \\ 9 \end{pmatrix}}}$$

*(c)*

$$\text{Cos } K\hat{A}L = \frac{\vec{AK} \cdot \vec{AL}}{|\vec{AK}||\vec{AL}|} = \frac{\begin{pmatrix} -5 \\ 5 \\ 11 \end{pmatrix} \cdot \begin{pmatrix} 2 \\ 4 \\ 9 \end{pmatrix}}{\sqrt{171} \cdot \sqrt{101}} = \frac{109}{\sqrt{171} \cdot \sqrt{101}}$$

$$\text{Cos } \hat{A} = 0{\cdot}829$$
$$K\hat{A}L = 33{\cdot}96$$
$$K\hat{A}L \doteqdot \underline{\underline{34^\circ}}$$

**Question 4**

*(a)* $y = 4x - x^2 = 0$ at P and O

$x(4 - x) = 0$

$x = 0 \quad x = 4 \Rightarrow \underline{\underline{P(4,\ 0)}}$

*(b)* $\left.\begin{matrix} R(0,2) \\ P(4,0) \end{matrix}\right\} \quad m_{PR} = \frac{2-0}{0-4} = \frac{2}{-4} = \frac{-1}{2}$

$$y = \frac{-1}{2}x + 2$$
$$2y = -x + 4$$
$$x + 2y = 4$$

Eqn$_{PR}$ is $\underline{\underline{x + 2y = 4}}$

*(c)* $y = 4x - x^2$

$2y = 8x - 2x^2 \quad$ AND $\quad 2y = -x + 4$

They meet where $8x - 2x^2 = -x + 4$

$$9x - 2x^2 - 4 = 0$$
$$2x^2 - 9x + 4 = 0$$
$$(2x-1)(x-4) = 0$$
$$2x - 1 = 0 \quad \text{OR} \quad x - 4 = 0$$
$$2x = 1 \qquad \underline{\underline{x = 4}}$$
$$\underline{\underline{x = \frac{1}{2}}}$$

At $x = \frac{1}{2}$; $y = 4\left(\frac{1}{2}\right) - \left(\frac{1}{2}\right)^2 = 1\frac{3}{4} \Rightarrow \underline{\underline{Q\left(\frac{1}{2},\ 1\frac{3}{4}\right)}}$

or $Q\left(\frac{1}{2},\ \frac{7}{4}\right)$

## Question 5

*(a)* (i) $L = 8x + 9y$

(ii)
$$8x + 9y = 360$$
$$9y = 360 - 8x$$
$$y = 40 - \frac{8}{9}x$$

$$\text{Area} = 6xy$$
$$= 6x\left(40 - \frac{8}{9}x\right)$$
$$\Rightarrow A(x) = \underline{\underline{240x - \frac{16}{3}x^2}}$$

*(b)*

$$A'(x) = 240 - \frac{32}{3}x = 0 \text{ at S.V.}$$
$$720 - 32x = 0$$
$$32x = 720$$
$$\underline{\underline{x = 22 \cdot 5}}$$
$$\underline{\underline{y = 20}}$$

| $x$ | | 22·5 | |
|---|---|---|---|
| $f'(x)$ | + | 0 | – |
| Shape | ↗ | → | ↘ |

Max T.P
(22·5, 2700)

$$\text{Max area} = 6xy = \underline{\underline{2700 \text{ m}^2}}$$

## Question 6

*(a)*
$$I_t = I_0 e^{-kt}$$
$$I_t = 120e^{-4k} = 90$$
$$e^{-4k} = 0 \cdot 75$$
$$-4k \ln e = \ln 0 \cdot 75$$
$$k = \frac{\ln 0 \cdot 75}{-4}$$
$$k = \underline{\underline{0 \cdot 0719}}$$

*(b)*
$$I_t = I_0 e^{-0.0719t}$$
$$I_t = I_0 e^{-0.0719 \times 10}$$
$$= I_0 e^{-0.719}$$
$$= 0{\cdot}487\, I_0 \Rightarrow 51{\cdot}3\% \text{ reduction}$$

---

**Question 7**

*(a)*

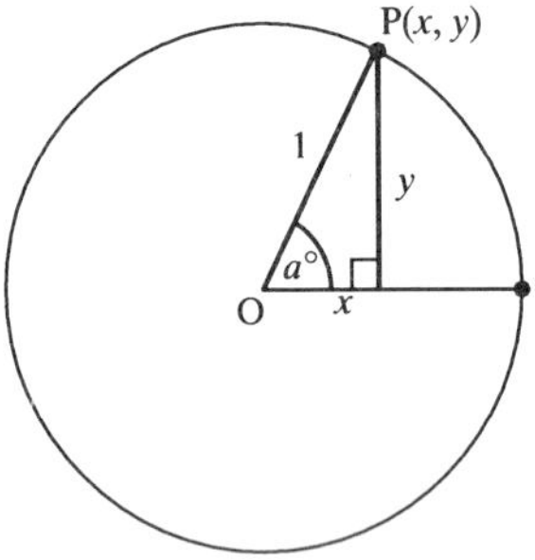

P(x, y)

1

y

$a°$

O

x

$$\left.\begin{array}{l} \sin a° = \dfrac{y}{1} \Rightarrow y = \sin a° \\ \cos a° = \dfrac{x}{1} \Rightarrow x = \cos a° \end{array}\right\} \Rightarrow \text{P}(\cos a°,\ \sin a°)$$

*(b)* Q $\left(\cos(a-45)°,\ \sin(a-45)°\right)$

*(c)* R $\left(\cos(a+45)°,\ \sin(a+45)°\right)$

*(d)*
$$m_{QR} = \frac{\sin(a-45)° - \sin(a+45)°}{\cos(a-45)° - \cos(a+45)°}$$
$$= \frac{\sin a\frac{1}{\sqrt{2}} - \cos a\frac{1}{\sqrt{2}} - \left(\sin a\frac{1}{\sqrt{2}} + \cos a\frac{1}{\sqrt{2}}\right)}{\cos a\frac{1}{\sqrt{2}} + \sin a\frac{1}{\sqrt{2}} - \left(\cos a\frac{1}{\sqrt{2}} - \sin a\frac{1}{\sqrt{2}}\right)}$$
$$= \frac{-\sqrt{2}\cos a}{\sqrt{2}\sin a} = \underline{\underline{\frac{-1}{\tan a°}}} \quad \left\{\text{or } \frac{-\cos a}{\sin a}\right\}$$

---

**Question 8**

$$\begin{aligned}
\text{Let }\ 2\sin x^\circ - 3\cos x^\circ &= k\cos(x-\alpha)^\circ \\
-3\cos x^\circ + 2\sin x^\circ &= k\cos x^\circ\cos\alpha^\circ + k\sin x^\circ\sin\alpha^\circ
\end{aligned}$$

Equate coefficients $\Rightarrow k\sin\alpha^\circ = 2$

$k\cos\alpha^\circ = -3$

$\Rightarrow \tan\alpha^\circ = \dfrac{-2}{3}$ Also $k^2 = 2^2 + (-3)^2 = 13$

$\alpha = \underline{\underline{146\cdot 3}}$ so $k = \underline{\underline{\sqrt{13}}}$

$$\begin{aligned}
\text{Hence }\ 2\sin x^\circ - 3\cos x^\circ = \sqrt{13}\ \cos(x-146\cdot 3) &= 2\cdot 5 \\
\cos(x-146\cdot 3) &= 0\cdot 693 \\
x - 146\cdot 3 &= 46\cdot 1,\ 313\cdot 9 \\
x &= 192\cdot 4,\ 460\cdot 2 \\
\Rightarrow \qquad x &= \underline{\underline{100\cdot 2,\ 192\cdot 4}} \quad \{\text{Subtract } 360^\circ\}
\end{aligned}$$

**Question 9**

*(a)* At P($p$, 4); $y = \dfrac{4}{x^2} \Rightarrow 4 = \dfrac{4}{p^2}$; At Q($q$, 1); $y = \dfrac{4}{x^2} \Rightarrow 1 = \dfrac{4}{q^2}$

$p^2 = 1$ $\qquad q^2 = 4$

$\underline{\underline{p = 1}}$ $\qquad \underline{\underline{q = 2}}$

*(b)*

$$\begin{aligned}
\text{Area} &= \int_0^1 \left\{x(x+3) - \left(x - \frac{x^2}{4}\right)\right\}dx + \int_1^2 \left\{\frac{4}{x^2} - \left(x - \frac{x^2}{4}\right)\right\}dx \\
&= \int_0^1 \left\{\frac{5}{4}x^2 + 2x\right\}dx + \int_1^2 \left\{4x^{-2} - x + \frac{x^2}{4}\right\}dx \\
&= \left[\frac{5x^3}{12} + x^2\right]_0^1 + \left[-4x^{-1} - \frac{x^2}{2} + \frac{x^3}{12}\right]_1^2 \\
&= \left(\frac{5}{12} + 1\right) + \left\{\left(-2 - 2 + \frac{2}{3}\right) - \left(-4 - \frac{1}{2} + \frac{1}{12}\right)\right\} \\
&= \frac{17}{12} + \frac{13}{12} \\
&= \underline{\underline{2\frac{1}{2}\ \text{units}^2}} \qquad \left\{\frac{30}{12}\ \text{units}^2\right\}
\end{aligned}$$

## QUESTION FREQUENCY CHART FOR HIGHER PAPERS

| TOPIC | Paper A | | Paper B | | Paper C | | Paper D | | Paper E | | Paper F | | Paper G | |
|---|---|---|---|---|---|---|---|---|---|---|---|---|---|---|
| | I | II | I | II | I | II | I | II | I | II | I | II | I | II |
| **Unit 1** | | | | | | | | | | | | | | |
| LO1 The Straight | 2 | | | 2 | 3 | 1 | | 7a | 1 | 1 | 1 | 3b | 2 | 1 |
| LO2 $f(x)$, Graphs, Solutions | 7 | 3 | 2,7 | | 5 | 4a | 4,10 | 2 | 2a | 5,6a,c | 5 | | 8 | |
| LO3 Basic differentation | | 1 | 9 | 7 | 4.7.11 | 2a,6 | 5,9 | 1 | 2,4,6 | 10 | 4a,9 | 2,3a,8 | 10 | 2,5 |
| LO4 Recurrence relations | | 5 | 6 | | | 3 | 6 | 5 | | 4 | 7 | 7 | 9 | |
| **Unit 2** | | | | | | | | | | | | | | |
| LO1 Factors, Remainder Theorem, Quadratic | 7 | 4 | | 1,8 | 1,10 | 2b | 3,7 | | 4,12 | 1 | 2,4b | 4a | 1 | 4 |
| LO2 Basic, Integration | 3 | 2 | 1 | 9 | | 8 | | 8a,b | 11 | 6b | 8 | 4b | 3 | 9 |
| LO3 Trig. form & equations | 5,9 | | 8 | | 6 | | 8 | 6 | 5 | 8 | 6,12 | | | 7 |
| LO4 Equation of circle | 4 | | 5 | 4 | | 7 | 1,12 | 7b,c | 8 | 2 | 3 | 5 | 4,5 | |
| **Unit 3** | | | | | | | | | | | | | | |
| LO1 Vectors | 1 | | 2,4 | 3 | | 5 | 2 | 3 | 9 | 3 | 11 | 1 | 8 | 3 |
| LO2 Further Calculus | 8,9,10 | 8 | 11 | | 2 | 9 | | 4 | 7,14 | 9 | 10 | | 11 | |
| LO3 Logs & exp. function | 11 | 7 | 10 | 6 | 8,9 | | 11 | | 3 | 7 | | 9 | 7 | 6 |
| LO4 Further Trig. & Wave function | | 6 | | 5 | 12 | 4b | | 8c | | 5 | | 6 | | 8 |

Printed by Bell & Bain, Ltd., Glasgow, Scotland, U.K.